AF759690

Springer Series on Cultural Computing

Editor-in-chief

Ernest Edmonds, University of Technology, Sydney, Australia

Series editors

Frieder Nake, University of Bremen, Bremen, Germany
Nick Bryan-Kinns, Queen Mary University of London, London, UK
Linda Candy, University of Technology, Sydney, Australia
David England, Liverpool John Moores University, Liverpool, UK
Andrew Hugill, De Montfort University, Leicester, UK
Shigeki Amitani, Adobe Systems Inc., Tokyo, Japan
Doug Riecken, Columbia University, New York, USA
Jonas Lowgren, Linköping University, Noorköping, Sweden

More information about this series at http://www.springer.com/series/10481

Christina Chau

Movement, Time, Technology, and Art

Springer

Christina Chau
Curtin University
Perth, WA
Australia

ISSN 2195-9056 ISSN 2195-9064 (electronic)
Springer Series on Cultural Computing
ISBN 978-981-10-4704-6 ISBN 978-981-10-4705-3 (eBook)
DOI 10.1007/978-981-10-4705-3

Library of Congress Control Number: 2017939618

© Springer Nature Singapore Pte Ltd. 2017
This work is subject to copyright. All rights are reserved by the Publisher, whether the whole or part of the material is concerned, specifically the rights of translation, reprinting, reuse of illustrations, recitation, broadcasting, reproduction on microfilms or in any other physical way, and transmission or information storage and retrieval, electronic adaptation, computer software, or by similar or dissimilar methodology now known or hereafter developed.
The use of general descriptive names, registered names, trademarks, service marks, etc. in this publication does not imply, even in the absence of a specific statement, that such names are exempt from the relevant protective laws and regulations and therefore free for general use.
The publisher, the authors and the editors are safe to assume that the advice and information in this book are believed to be true and accurate at the date of publication. Neither the publisher nor the authors or the editors give a warranty, express or implied, with respect to the material contained herein or for any errors or omissions that may have been made. The publisher remains neutral with regard to jurisdictional claims in published maps and institutional affiliations.

Printed on acid-free paper

This Springer imprint is published by Springer Nature
The registered company is Springer Nature Singapore Pte Ltd.
The registered company address is: 152 Beach Road, #21-01/04 Gateway East, Singapore 189721, Singapore

Contents

Chapter 1
Introduction: Peculiar Time

Abstract This chapter highlights that kinetic sculpture has at times been used in art history today to refer to an antiquated artistic practice, which has consequently contributed to a long-standing 'flagrant dearth' of critical and historical engagement with movement in art. Despite this attitude there has been an increasing amount of exhibitions which have turned to avant-garde kinetic artworks and sought new interpretations of the roles and affects of movement in sculpture and installation. This chapter argues that this turn towards movement in art contributes to a key trend in contemporary art and give indication to our current sense of contemporaneity in art and society.

1.1 Movement and Technology for Contemporary Audiences

For contemporary audiences the notion of technology in art feels like a given. In many situations the experience of art is dependent on the performance of technology that the audience interacts and moves with the work of art. We're accustomed to an array of technological spectacles that are art dynamic, including immersive installations, interactive and participatory digital works, advanced robotics, carnival-esque light shows, and steampunk performances filling art institutions. While some argue that technological media in art is often marginalised from discussion around traditional visual arts, there is undoubtedly an increased use of mechanical and digital technology in museums, galleries, biennales and studios today.

Movement, Time, Technology, and Art focuses on the use of technology in art in a specific way, by concentrating on the affects of technology to create actual of movement in art. Actual movement (or kinesis) has the ability to affect our perceptions of the present in ways different to representations of movement in the screen arts. For instance, movement has a powerful capacity to generate suspense, protract a long duration, or cut our experiences up in multiple instances as we move-with the artwork. Actual movement can also raise questions around our interactions with an artwork, how we could respond to its movement and how we

© Springer Nature Singapore Pte Ltd. 2017

C. Chau, *Movement, Time, Technology, and Art*, Springer Series on Cultural Computing, DOI 10.1007/978-981-10-4705-3_1

might move our body around or in relation to it. Rather than looking at *how* technology has been used in art, this book focuses on what *affect* movement in art has on use as audience members, and how these affects have been framed over time in art history.

1.2 What is Kinetic Sculpture and Kineticism?

The term 'kinetic art' in art history and theory usually refers to the work of avant-garde artists who were interested in creating movement with mechanical media. However over time kinetic art has also accumulated some critical baggage and has been considered a forgotten and antiquated experimentation between with movement, technology and art. For example, the exhibitions *Force Fields, Phases of the Kinetic* (2000) was publicised as 'an investigation of movement in art, which in the mid-twentieth century became obscured by waves of more fashionable movements' (Nash 2000, p. 313). Similarly the exhibition *Ghosts in the Machine* at the New Museum in New York (2012) was promoted as 'an unsystematic archive' that displayed a 'cabinet of curiosities' made by artists of past technological ages, predominantly mechanical sculptures that move (*New Museum*). In these instances kineticism is approached as the aestheticisation of modern industrial technology in movement, and as a tendency that reached a climax in popularity during the 1960s, and is now remembered as a nostalgic trend that is 'rooted in another age' (Riding), that came and went quicker than a 'flash in the pan' in the 1960s (Bois 2000, p. 145).

Even though the use of actual movement is common to a variety of contemporary art practices, the terms 'kinesis', 'kinetic', and 'kineticism' have been used to describe the movement of industrial machines in art that act as an art historical precursor to the 'more refined' contemporary media art practices (Popper 2007). Therefore there is a history that assumes mechanical movement in art as the clunky ancestor of the sleek and immediate digital technology that we often see in contemporary art.

The assumption that kinetic art is an obsolete modern antiquated practice suggests that we've neatly categorised the practice, and have also moved on to more advanced and experiments between art and technology. This attitude is partially due to kineticism being a form of technological art, which according to Edward Shanken "little scholarship has explored the relationship between technology and conceptual art" (2004). Consequently Arnauld Pierre has argued that there is a long-standing 'flagrant dearth' of critical and historical engagement with kinetic art (p. 91).

Why would there be a deficiency in discussing the movement of technology in art especially when movement is commonly performed in contemporary art? One of the first art historians to consider kineticism as an out dated practice was Jack Burnham in *Beyond Modern Sculpture: The Effects of Science and Technology on the Sculpture of this Century*. Burnham argued that prior to 1968 kineticism had the potential to become a mainstream artform because it intersected with science, art and technology, but failed to do so because many of them did not keep pushing the boundaries of what technology could do at the time, and focused more on artistic experimentations (pp. 218–221).

Even though Burnham retired as an art critic and historian shortly after the publication of *Beyond Modern Sculpture*, his arguments in relation to art, science, and technology have remained influential in the field of media art history and contemporary art criticism (Shanken; Terranova). According Caroline Jones in *Artforum*, Burnham's work has come 'to define some of the most significant cultural developments of our time' that 'suffuses the art world as we know it' (Jones 2012, pp. 113–114). Jones' comment indicates that Burnham's approach to modern sculpture has influenced a contemporary understanding of kineticism, and also the wider milieu of contemporary art theory and history.

Art historians such as Jeremy Benthall and Frank Popper have expressed similar attitudes toward kinetic art to Burnham, although they were more sensitive to the effects of technology in movement in subsequent media art practices. In 1972 Benthall commented that kineticism inevitably would not continue, but would become a tendency from which new practices would be borne. Additionally, Popper has argued that kinetic art is predominantly remembered as an early mechanical precursor to 'more refined' digital and contemporary technological art practices (2007). For Popper, kinetic art is centred on the aesthetic of 'movement expressed by movement itself…as an attempt to incorporate the notion of space-time in the plastic work' (p. 221), and is considered as an early mechanical experiment that became absorbed by digital art practices such as virtual art. Popper's *Origins and Developments of Kinetic Art* remains the most comprehensive analysis of kineticism in contemporary art history to date and many contemporary scholars such as Shanken (2003) have since treated Popper as a pioneering historian in the field of art and technology (Rush 1999, p. 172).

Even though the term 'kinesis' refers to any form of dynamic movement of any media technology, Burnham, Benthall, and Popper consider kinetic art as an orchestration of movement by modern mechanical media and is art historical precursor contemporary media art practices (Popper 2007, p. 1; Burnham 1968, p. 219). This distinction therefore restricts kineticism to an older mechanical age, and is a key contributing reason why kinetic art has been associated as an antiquated aesthetic that has not often continued through to discourses of contemporary art and technology.

These assumptions are often drawn in art history to build a genealogy of the influences and artistic milestones that chronologically lead toward the contemporary milieu of media art. The attitudes projected by Benthall, Burnham, and Popper, have been influential in contemporary media art history, and raise a number of questions regarding the roles and effects of contemporary kinetic art history, which this book moves from.

1.3 Exhibiting Kinetic Art

A number of exhibitions in the early 2000s echoed Burnham's approach by framing kinetic art as a marginalised and forgotten artistic experiment with mechanical media. At times these exhibitions used the term 'kinetic' to refer to a forgotten historical practice that is 'rooted in another age' (Riding). In some situations, such as *Force Fields: Phases of the Kinetic,* the curator, Guy Brett, returned audiences to a 'forgotten' preoccupation with movement by modern artists, in order to reflect on the changing conceptions of movement, time and perception (Nash 2000, pp. 313–316). Shown at the Museu d'Art Contemporani de Barcelona (MACBA) and the Hayward Gallery in London, Brett aimed to reintroduce audiences to avant-garde kineticism through *Force Fields*. The exhibition included early European kinetic sculptures such as Marcel Duchamp and Many Ray's *Rotary Glass Plates [Precision Optics]* (1920), László Moholy-Nagy's *Light Space Modulator* (1922–1930), as well as kinetic works by Alexander Calder, Sol Le Witt, Len Lye, Hans Haacke, Jean Tinguely, Lygia Clark, Hélio Oititica, Julio le Parc, and David Medalla. According to Brett, artists such as these were included to 'reintroduce us to an investigation of movement in art which in the mid twentieth century became obscured' (Nash 2000, p. 313), and are at times thought of as forgotten misfits (Borja-Villel 2000, p. 7).

Other exhibitions that returned to earlier avant-garde kinetic experiments include *Shakin' The Contemporary Kinetic Aesthetic* (2012), at the Gold Coast City Gallery, Australia; *The Pleasure of Light: György Kepes and Frank Malina at the Intersection of Science and Art* (2010), at the Ludwig Museum in Budapest; *Zero: Artists of a European Movement* (2006), at the Museum der Moderne in Salzburg; *Beyond Geometry: Experiments in Form 1940s to 1970s* (2004)at the Los Angeles County Museum of Art (LACMA); as well as *Geometry of Motion: 1920s/1970s* (2008), at MoMA, New York, and *Luce e Movimento* (2010) at the Signum Foundation, Venice. These exhibitions were each publicised as an homage to early avant-garde kinetic artists. While *Luce e Movimento* paid particular attention to the works and artists that exhibited with the Denise René gallery in Paris during the 1950s and 1960s, *Geometry of Motion* focused on what the curator described as artists of a 'neue optik' or 'new vision' in the 1920s and situational aesthetic artists of the 1970s.

Ghosts in the Machine at the New Museum in New York exhibited a variety of contemporary and historical artworks that aimed to investigate artists interested in mechanical, optical, and virtual art. The exhibition was interpreted by some as an historical exhibition that was 'a little short on living, breathing artworks, and slightly overloaded with rather stale ones and other objects and diagrams that, altogether, function primarily as interesting period pieces or historical artifacts' (R Smith). The exhibition included a strong focus on kinetic artworks to represent 'a prehistory of the digital age' to be considered 'a cabinet of curiosities' (Schjeeldahl), full of 'neglected objects' (R Smith).

However the initial impetus for this book began with a number of recent exhibitions that have regarded avant-garde kinetic art as a means for reflecting on the changing conceptions and expressions of temporality over time. More specifically *Movement, Time, Technology, and Art* is centered on the use of movement in contemporary art history as an instrument to express and orchestrate new perceptions of temporality with mechanical and digital media.

Many exhibitions have broken away from the assumption that kinetic art is an antiquated practice, and revisited key kinetic artworks to reflect on modern and contemporary approaches to movement, time and technology in art. For instance, *Vibration, Vibração, Vibración: Latin American Kinetic Art of the 1960s and 70s*, at the Power Collection, University of Sydney (2012), incorporated a number of artworks that have influenced contemporary interpretations of kineticism. Curated by Susan Best, the exhibition presented a renegotiation of Latin American kinetic art that ran parallel to dominant Western tendencies in contemporary art history. For Best, avant-garde Latin American kinetic art underwent patterns of 'survival and revival' that did not refer to the dominant reception of kineticism in Europe and North America. Best's curatorial selection saw these artists as participating in 'alter-forms' and tendencies related to movement. Best's approach contradicts Pierre's claim that there is a current absence of critical and historical studies of kinetic art, because *Vibration, Vibração, Vibración* offers a renewed historical perspective of the role and effect of kinetic form.

Similarly, *Points of Contact* (2010) at the Govett Brewster Gallery in New Plymouth, New Zealand, incorporated works by Hélio Oticica, Len Lye and Jim Allen to reconsider connections between kinetic and conceptual art in the country where post-object tendencies 'unlike its American counterpart derived not from minimalism' in New Zealand. *Points of Contact* instead considered the idea that conceptual art emerged from kinetic sculpture and kinetic environments rather than the other way around,

The exhibition *Moving Parts: Forms of the Kinetic* at the Graz Kunsthaus (2004) attempted to draw stronger connections between kinetic and contemporary sculpture in order to reflect how the experience of mechanical movement has 'come to appear as everyday normality to us' (Magnaguagno). The exhibition focused on early kinetic artists such as Jean Tinguely, Hans Haacke, George Rickey, and Julio le Parc, alongside contemporary kinetic works by Olafur Eliasson, Rebecca Horn, and Jeppe Hein, to highlight 'a renewed interest in the possibilities of kinetic forms' (Pakesch). The exhibition also emphasised the necessity to reflect on modern kinetic works in relation to contemporary pieces in order have 'a better understanding of current approaches' to movement, time and the machine aesthetic in contemporary art (Pakesch p. 17).

Additionally, *Under Destruction* (2010) at Museum Tinguely in Basel was a commemorative exhibition for Tinguely's auto-destructive work, *Homage to New York* (1960). The exhibition incorporated emerging contemporary artists that continued Tinguely's approaches to movement and time in contemporary contexts. Finally, several biennales, art fairs, and symposia have emphasised a desire to renegotiate the contemporary art history of kineticism. This is inclusive of the annual art fair orchestrated by the Kinetica Museum since 2009, and the 2013 International Kinetic Art Exhibit and Symposium at Boynton Beach in Florida.

These exhibitions sought new ways of thinking about modern kinetic art in order to discuss new emerging relationships to movement and time in contemporary society. The exhibitions are also indicative of what art historian Edward Shanken has recently identified; that the traditional forms of kinesis from the early modern avant-garde have 'increasingly become incorporated into contemporary art practices' (Shanken 2012). The exhibitions that I have addressed are evidence of this incorporation on an institutional level: they indicate that new methods for interpreting, recalling and historicising kinetic sculpture have broken away from the assumption that kinetcism is antiquated and begun to incorporate the practice within the frameworks of contemporary art and art history. They also demonstrate that the dearth in critical and historical engagement with kineticism is undergoing a process of review, and new understandings between avant-garde kinesis and contemporary art are emerging (p. 91).

1.4 Time, Contemporaneity, and Kinetic Art

While all works of art present a relation to time in some way, kinetic art is a unique orchestration of movement that unfolds in duration with the viewer, and therefore enables an expression of time in a way that is specific to the medium. To experience a kinetic artwork one must take time and move with it in duration, to (potentially) perceive movement in time in new ways. This book looks at what it means to take time with a kinetic art work, and points to some of the affects of a moving artwork. Therefore rather than asking *how* does one make a moving artwork with technology, this book unravels some of the ways movement in art can alter our perception of time through movement.

Time has been connected to kinetic art throughout the twentieth century. From as early as 1919 Russian constructivist Naum Gabo called upon his audiences to focus on the present temporality and break away from traditional modes of producing and consuming art. Gabo explicitly expressed this urge through his kinetic sculpture *Kinetic Construction: Standing Wave*, and his *Realistic Manifesto*, where actual movement (rather than representations of it) was seen as a revolutionary form of art:

> The attempts of the Cubists and the Futurists to lift the visual arts from the bogs of the past have led to only new delusions. ...We assert that the shouts about the future are for us the same as the tears about the past: a renovated day-dream of the romantics...Today we take the present (Danchev Danchev 2011, p. 93).

In the late 1950s and early 1960s Jean Tinguely created auto-destructive mechanical landscapes and urged people to '[l]ive in the present, live once more in time and by Time [sic]—for a wonderful and absolute reality...Stop painting "time"...live in time and according to time for a wonderful and absolute reality'(in Danchev 2011, pp. 336–337). More recently, contemporary artists such as Olafur Eliasson have also used movement in installation art in order to discuss the subjective perception of 'now' and the effects that movement gives and enables audiences to ‘perceive yourself perceiving’ in the present temporality (Eliasson 2007, p. 55).

Attention to movement and time produces a number of questions around the changing interpretations of modern kinetic art and its relationship to contemporary art, particularly in the areas of conceptual, participatory art, and the machine aesthetic. While many others have analysed the changing conceptions of time through art and media such as Sven Lütticken’s *History in Motion: Time in the Age of the Moving Image*, this book reevaluates the relationship between kinetic art and time. This reconsideration also opens new avenues for understanding the roles and effects of kineticism in art today that intersect with contemporary approaches to mechanical, and digital media arts. Therefore, rather than approaching kinetic art as an predecessor to media art today, this book is focused on kinetic art as a tool for exploring the nature of contemporaneity over time.

It is apparent, then, that there is a tension between two broad approaches to kinetic sculpture in contemporary art today. While Burnham and Popper’s scholarship have been influential criticisms of kineticism in relation to media art practices today, there is also a clear contemporary tendency to incorporate kinetic art into discussions that reflect on the nature of art and society today. This tension points to a number of questions around the changing roles and effects of kinesis in art, and suggests that a further analysis of kineticism and its relationship to contemporaneity is needed in order to further understand aspects of contemporary art.

1.5 What is Contemporary Art?

Thinking about the roles and effects of kinetic art in society today is particularly complicated when the term ‘contemporary art’ is an elusive term. As expressed by Hal Foster in ‘Questionnaire on “The Contemporary”’ recently in *October*, an understanding of contemporary art today is under defined. Foster explains that contemporary art today:

> [S]eems to float free of historical determination, conceptual definition, and critical judgment. Such paradigms as ‘the neo-avant-garde” and “postmodernism,’ which once oriented some art and theory, have run into the sand, and, arguable, no models of much explanatory reach or intellectual force have risen in their stead. At the same time, perhaps paradoxically, ‘contemporary art’ has become an institutional object in its own right…most tend to treat it as apart not only from prewar practice but from most postwar practice as well (1999, p. 3).

In light of this open and under formed understanding of the characteristics of contemporary art, many scholars have attempted to clarify the ambiguous, re-forming and complex state of art today. For instance, Terry Smith’s *What is Contemporary Art?* proposes a method for unraveling some of the key tendencies that have emerged in art since the early 1990s. What is most useful is Smith’s interest in the expressions of contemporaneity today. Smith delineates three general and interweaving currents that have formed since the early 1990s.

For Smith, One of these tendencies points to artists and institutions that recall and resensationalise modern approaches to art that contributes to a specific characteristic of contemporary art today. The extent to which contemporary art sensationalises modern tropes of art also suggests that contemporary society is an extension of modern society rather than a tendency that returns to or refashions an historical modernity. While postmodernity is at times regarded as a chronological and theoretical addition to modernity, it is also considered, as Frederic Jameson explains, to be a distinct framework that destabilised the power structures laden within modernity (1991, pp. 56–61), leaving contemporary art and theory in a posthistorical state. Contrary to this, Jameson has recently argued that contemporary society expresses a number of modern processes that exist within it (2003, pp. 717–718). As with Jürgen Habermas (1981) who understood modernity as a system that is endless in nature, Jameson argues that the themes of universality, progress, colonisation and reproduction continue in the present day, suggesting that contemporary art and society are consequently characterised by modern aims.

According to Smith, a number of artists and institutions today pursue a modern tradition of seeking newness by creating art that produces new effects. They experiment with the formal traditional expectations of art and attempt to confront modes of representation (2008). These artists and institutions 'pursue the key drivers of modernist art: reflexivity and avant-garde experimentality' (2009, p. 265), and political contexts, and depend on modern avant-garde tendencies in order to elaborate on contemporary society. This tendency is what Smith refers to as 're-modernism'; and, at times, he uses 'resensationalism' to refer to artists and institutions that perpetuate and re-form modern approaches to creating and interpreting art. This is inclusive of Richard Serra, Julien Schnabel, the Young British Artists (yBAs), Matthew Barney, Cai Guo-Qiang, Gerhard Richter and Jeff Wall and institutions such as DIA Foundation, Tate Modern and the refurbished Museum of Modern Art in New York. While Smith's argument is specific to these artists and institutions, his method for approaching contemporary art today raises questions around further aspects of remodernism at play.

There are times when modern kinetic art has been resensationalised by contemporary artists and institutions in order to discuss, express and orchestrate old and new perceptions of temporality through art in order to further understand the nature of contemporaneity in society today. Smith's approach to contemporary art forms a critical basis for this book because of his emphasis on modern tropes in art at play within the schema of contemporary art and contemporaneity. What is most useful to Smith's understanding, is that it can be used as an opening for considering the role and effects of avant-garde kineticism and relationships between kinesis and contemporaneity.

Smith is not alone in identifying modern trends within the milieu of contemporary art. Nicolas Bourriaud argues, for example, that contemporary art is embarking on a phase of 'altermodernism' and considers contemporary art as a new form of modernity. In 2009 Bourriaud curated the third Tate Triennial titled *Altermodern*, and proposed that the artists explore the current globalised world as a form of new universalism. However, unlike Smith's description of an antonymous, plural, evasive, and contradictory contemporary art, Bourriaud focuses on the themes of global society, migration, multiculturalism, identity and communication to form a new singular modernity expressed through contemporary art.

These observations also indicate that a wider social and political conception of temporality is in a state of transition, and it is this state of transition that has come to characterise contemporary art. Considering this, kinetic movement functions more than a mechanical art, but can also be used to explore our relationship to time in current society. This is particularly prevalent when considering Pamela Lee's argument, in *Chronophobia* that contemporary society since the 1960s has tended to produce phobic expressions of time, as well as Erin Manning's study of movement that moves from a state of preacceleration (pp. 5–7). Lee and Manning not only analyse motion, actual movement, and time-based arts, but they also consider contemporaneity to be a specific engagement with being with time in art today. Lee has argued that artists in the 1960s have enframed contemporary expressions and reactions to self-reflexive engagements with temporality in art, much of which depended on the experimentation of technology in art.

Manning's approach to incipient movement articulates the immanence of movement moving (54). She addresses preacceleration through various practices to indicate that there is a strong exploration of movement by artists, for which conceptual language lacks a modality for interpreting. Her attempt to fill part of this void is done so from the perspective that depends on a concept of thought in motion, or writing with the 'force of movement moving' (p. 3). Unlike Lee, Manning has resisted a chronophobic reading of time-based arts that understands time and movement predominantly as tools for displacement by modern and contemporary artists.

Smith, Bourriaud, Lee, and Manning each provide commentary on the tendencies and characteristics of contemporary art, as well as society's changing relationship with temporality. Their approaches also informs key critical aspects to my approach to kinetic art and art history today. Even though there are many differences among their arguments, they form contributions to a larger discussion of the peculiar relationship to the present temporality in art today. Manning and Lee have drawn from modern philosophers like Bergson and Whitehead to frame contemporary understandings of temporality. While Bourriaud and Smith have significant differences in their arguments regarding how modernity persists in contemporary art and identity, they also recognise that contemporary artists are perpetuating modern themes within their artworks. Additionally, these approaches are helpful for engaging with actual movement in contemporary art and art history and the affects that movement has on the view in relation to the perception of temporality and contemporaneity.

1.6 Talking about time

Modern philosophy often approaches temporality as an absolute form, or what Bergson regarded as the Whole *duree* (1911). By contrast, postmodern art and theory can broadly be understood to have destabilised the power structures laden within modernity through a system of reterritorialisation of temporal and spatial codes. According to Jameson this was executed to the degree that 'time has become a nonperson and people stopped writing about it' (2003, p. 695). This conjures the 'end to temporality' through cessation of time as a philosophical tool to understand society because people no longer visualised the world from a temporal point of view. Jameson emphasised that since the industrial revolution modern society has positioned time-space in 'homeric opposition' to one another, rather than progressing in relative synthesis (2003, p. 698). For example, film and photography produced representations of time to be perceived from a spatial point of view and rendered time into isolated segments of space. Although, as Deleuze suggested, early modern cinema produced a movement-image that represented time from a spatial point of view (1986), just as modern cinema in the mid twentieth century began to be created to explore movement from a temporal vantage point (1989).

These changing approaches and conceptualisations of temporality indicate that each society expresses a specific contemporaneity, i.e., a specific relationship to being in and with time. As Althusser and Balibar have suggested, each society visualises, interprets, and understands temporality in ways that are specific to its own social context: each society has their own *peculiar time*. If each period builds its own specific interpretation of time, the transition from late-modernity to post-modernity is generally described as a process from time's 'livid final flame' towards a de-emphasis or disregard for temporality (Lefebvre 1991; Soja 1989; Foucault 1986) For Jameson the 'end of temporality' became a catchphrase for postmodern theory because 'it was widely rumored that space was supposed to replace time in the general ontological scheme of things. At the very least, time had become a nonperson' (1991, p. 695). Therefore while the *peculiar time* of contemporary society is characterised by Jameson as a zone of non-temporality, Smith draws from contemporary artists and institutions to demonstrate that time is no longer necessarily a function of speed, but can also be a mechanism to navigate and assert the peculiar nature of contemporaneity in contemporary society.

If contemporary art and society is in a state of transition, perhaps this very condition has come to be a key defining characteristic of contemporary art and society. Even though the term 'contemporary art' is not a neologism, there have been a substantial number of scholars and curators who have begun to analyse contemporary art as an emerging tendency that is informed by, and yet defers to, modern and postmodern art. This suggests that a new peculiar interpretation of time is emerging in both art and theory, which is specific to contemporary society. For instance, previous descriptions of time (Kubler 1962), such as speed, acceleration and the instant (Virilio 1984; Derrida 1984), still exist within the contemporary, but no longer serve explicitly as descriptions of its peculiar time (Gere 2006; Grosz 2004).

For Jameson (2003) the key point of difference that the present day has from modernity and postmodernity is that contemporary globalisation reconciles time and space to no longer be in mutual opposition to one another. This observation also provides an opening for considering a renewed investment in temporality and movement in art.

In art today there is a coexistence of distinct but simultaneous temporalities. For Smith in contemporary society there are 'different ways of *being* in relation to time, experienced in the midst of a growing sense that many kinds of time are running out…what is it to be *with* time, to be contemporary' (2009, pp. 3–4). Smith creates an elusive and contrary temporal landscape. He continues:

> the current situation…is characterised more by the insistent presentness of multiple, often incompatible temporalities accompanied by the failure of all candidates that seek to provide *the* overriding temporal framework—be it modern, historical, spiritual, evolutionary, geological, scientific, globalizing, planetary…Everything about time these days—and therefore about place, subjectivity, and sociality—is at once intensely *here*, is slipping, or has become artefactual (p. 196).

For Smith, time today is addressed as evasive, contradictory and antonymous in nature with a sense of urgency, in order to understand the 'deepest sense of the contemporary: what it is to be *with* time, to be contemporary' (pp. 3–4). The present does not move forward or backwards but is expressed in a contradictory and unpredictable manner in contemporary art. This uncertainty about how time is mediated not only highlights the contrary, elusive nature of temporality, but also heightens our awareness of this ambiguity as a defining characteristic of time in contemporary art. Consequently contemporary artists have become attuned to, and affected by expressions of multiple and, at times, conflicting time scales; simultaneously geological, modern, historical, spiritual, global, scientific and cosmological time scales. This signifies not only an expansion of temporal experiences but also an ambiguity as to how they contribute to a defining character of contemporary time.

Temporal uncertainty does not necessarily suspend us in time, or accelerate its speed, but heightens our perceptual sensitivity to it as constituting multiple, incompatible temporal modes. Conflicting approaches to time all understand the present to function differently. For instance, the temporality that is expressed and encouraged through remodernism is one of many simultaneous contemporary time codes that exist in art today. Consequently, the re-exhibition of kinetic art functions in two ways: to recall modern avant-garde conceptions of temporality and to encourage discussions about how current society has deferred to previous modern temporalities.

The negotiation of temporality in contemporary art is also a key area for understanding the nature of contemporaneity in the present day. For instance, exhibitions such as *Where Are We Going?* at the Palazzo Grassi in Venice in 2006 returned to modern questions of time and space to investigate the nature of contemporaneity in this social climate. A key work at the exhibition was Damien Hirst's *Where Are We going? Where Do We Come From? Is There a Reason?*

which proposes questions around how contemporary society understands and expresses the concepts of time. The work consists of a stainless steel and glass cabinet installed in a T-section that displays a variety of animal skeletons. If the decades after 2000 can be described as being after 'the end of history' and after 'the end of temporality', Hirst's work is an invitation to reconsider and relocate the self in time and space and is but one example of contemporary artists grappling with this polemic. This is an articulation of an ongoing curiosity about time by recycling and referencing Paul Gauguin's modern colonial works, thereby reaffirming Smith's argument that contemporary artists such as Hirst perpetuate modern avant-garde practices.

Additionally there have been a number of contemporary exhibitions that have turned towards temporality as a means to reflect and explore contemporary identity. These include; *As Soon as Possible* at the Palazzo Strozzi, Florence (2010); The Whitney Biennale (2008) which projected the theme 'temporal tenuousness—the sense many artists feel of being in translation' (Momin); *Time Change*, at the Whitney Biennial (2008); *On Time. Disarming Matter* at the The Courtauld Institute of Art's East Wing VIII exhibition; the 2012 Kinetica Artfair themed 'Time, Transformation and Energy'; and the symposia series, *Time, Space and Energy* Symposium at Arti et Amicitiae, Amsterdam (2007), which debated the strong presence of time and temporality as an artistic tool in contemporary art. These exhibitions show contemporary artists that are interested in expressing temporality in new ways, in order to investigate how time is conceptualised in contemporary society in a manner that is inherent to current social and political settings. These exhibitions indicate that there is a continued desire to discuss and express the role of time in art in order to explore various motifs of temporality specific to contemporary society.

1.7 Chapter Outline

Sydney's sixteenth Biennale, *Revolutions—Forms that Turn* in 2008 is a significant exhibition in this book is used to raise a number of key research questions that underpin my understanding of contemporary approaches to kinetic art history. The book opens with an analysis of how kinetic sculpture can be incorporated into an art historical framework that sensationalises modernity within contemporary art. Unlike other biennales that often engage with the emerging local and global currents of contemporary art, *Revolutions—Forms that Turn* was contemporary art historical reflection on the presence and effects of the modern avant-garde in art today. Through Terry Smith's method for approaching and analysing contemporary art I argue that modern avant-garde kinetic sculpture is used in *Revolutions—Forms that Turn* to inform audiences of a peculiar consciousness of temporality. I also argue that contemporary art history incorporates modern avant-garde engagements with time and movement to reflect on the changing expressions of the present temporality, over time.

The avant-garde artists that feature in the biennale are seminal expressions of temporality. Combined, they create a critique of a modern machine aesthetic, and provide a self-reflexive approach to destabilising traditional modes of representation. Furthermore, the presentation of contemporary kinetic artworks alongside their historical antecedents draws a link between contemporary kineticism and early European experiments with mechanical movement .

For Carolyn Christo-Bakargiev, the artistic director of the biennale, the avant-garde artworks that featured in the exhibition were overall considered as being at once historical and contemporary. This is because, according to Christov-Bakargiev, contemporary art history draws from, and coalesces the recent past in order to understand the nature of contemporaneity today (2008, pp. 30–33). If this is the case, then not only is avant-garde kinetic sculpture considered to be within the broad setting of contemporary art, but also the artists and the artworks of this period can be used to understand the changing conceptions of time in contemporary art history, as well as being a resource for understanding aspects of art and society today.

Revolutions-Forms that Turn also focused on the changing conceptions of temporality in society and used a number of avant-garde kinetic artworks to help facilitate out this discussion. For Christov-Bakargiev, contemporary history is considered a coagulation of temporalities that the contemporary draws from to inform, reflect and express identity in the present day. For Christov Bakargiev's 'everything that exists in the world is of my time, whether it is an old 1950s Bakelite telephone, or an artwork made two years ago or today, simply because it is *cum-loco*—with place' (p. 33): That is, what is remembered by the present day is contemporary. In this regard avant-garde kinetic sculpture is not necessarily an obscured forgotten practice, but a resource to consider the role and presence of modern avant-garde art as a feature of contemporary art history. The Biennale has been discussed as an art historical lesson for viewers to posit modern avant-garde artists and their preoccupation and expression of temporality as key aspects of contemporary art history (Desmond 2008). If this is the case, Christov-Bakargiev attempts to pinpoint how society today expresses a specific consciousness of time.

Considering the tendency to discuss and contribute to understandings of contemporaneity in contemporary art, it is therefore no surprise that exhibitions such as *Revolutions—Forms that Turn* are curated with a renewed interest in avant-garde kinetic art. This is because, as I will argue that kinetic art and art history is a key resource for mapping the changing conceptions of contemporaneity. Additionally, because contemporary artists and institutions often recall the recent past to inform characteristics of contemporaneity, the study of avant-garde kinetic art is also valuable for understanding the peculiar time unique to contemporary contexts.

Chapter 2 focuses on the influences of this relationship to time in contemporary art. In a variety of respects, this relationship to temporality in art is borne from a turbulent relationship with temporality in Europe and North America in the 1960s. Michael Fried's seminal essay, 'Art and Objecthood' (1968), the exhibition *Directions in Kinetic Art* curated by Peter Selz, and the auto destructive kinetic sculptures by Jean Tinguely, *Homage to New York* (1960) and *Study for an End of the World* (1963/1964), provide a number of conflicting approaches to duration in art. The approaches to time presented by Fried, Selz, and Tinguely react to the role and effect of duration in art in conflicting and disparate ways, and culminate to signify a larger uncertainty towards how time is conceptualised in a period of technological transition.

I argue that artists such as Tinguely attempted to expand the ways in which time could be interpreted through movement . Tinguely used the movement of mechanical, televisual, and electronic media to emphasise that these technologies move differently and give rise to new temporal systems. Tinguely's sensitivity towards the different, and at times conflicting rationalisations of time through different mechanical media attempted to expand on the conceptualisation of time through movement in art by bringing attention to these differences.

In Chapters 3 and 4 I further draw out a number of tensions in relation to movement , time and art in the 1960s and early 1970s. While the 1960s has been described as the decade of the 'kinetic kraze', the technological turbulence during this time was overcome by an attempt to use spatial concepts to address the role of duration in art. One key example of this that greatly affected the creation and interpretation of kinetic art is Jack Burnham's argument in 'Systems Esthetics' (1968b). Chapter 3 addresses Burnham's appropriation of Ludwig von Bertalanffy's General Systems Theory as a method for engaging with emerging artistic practices at the time.

Burnham's approach to systems theory in art developed with, and alongside, German-born artist Hans Haacke. The artist's interest in incorporating environmental systems in his art is one of Burnham's earlier examples of systems aesthetics at play. According to Burnham, Haacke's work highlights a significant shift in art that moves from technological art towards the orchestration of cybernetic systems (Burnham 1969). However, Burnham's argument is dependent on the misdirection of Haacke's early systems art in order to affirm that kinetic artists remained unrequited by their aims, and failed to remain relevant to discussing art from a systematic point of view. Burnham was not alone in his disappointment in experiments with art and technology, as Garnet Hertz has recently reflected it "seemed like the whole art and technology movement of the late 1960s seemed to lose some of its initial momentum" (quoted in Candy and Edmonds p. 9).

In deference to Burnham's premise, actual movement is also useful for elaborating on perception as a systematic process because the arrangement of actual movement in Haacke's work is also a central aspect of systems aesthetics in art and connects with the material and immaterial implications of systems theory. As Haacke explains:

> A 'sculpture' that physically reacts to its environment is no longer to be regarded as an object. The range of outside factors affecting it, as well as its own radius of action, reach beyond the space it materially occupies. It thus merges with the environment in a relationship that is better understood as a 'system' of interdependent processes (in Burnham 1974, p. 22).

Burnham's approach to systems theory has been greatly influential in contemporary interpretations to systems theory in contemporary art and art history today. For instance Caroline Jones has suggested that Burnham's "systems thinking now suffuses the art world as we know it" (2012). While Jones has suggested that Burnham's influence has been broad reaching, others such as Ernest Edmonds have recognised the influence systems theory and Jack Burnham have had on contemporary interactive, time-based and digital art (pp. 18–24).

While Burnham used systems theory to reconsider approaches to art Frank Popper explored the effects of the emerging post-object aesthetic in relation to kinetic art. Chapter 4 addresses Popper's argument in *Art, Action and Participation* that kinetic art in the 1960s and early 1970s became increasingly dematerialised to the point of entire invisibility. This dematerialised work of art points to a de-emphasis of form, and the prioritisation of the concept of an artwork, and led to what can be described as a post-object, or post-formalist aesthetic. For Popper, demateriality is the catalyst for entirely disregarding the form of movement, and from which artists became engaged with relational and communicative movements (1975, pp. 7–12). Popper in *Art, Action and Participation* and, later, *From Technological to Virtual Art*, also delivers a digitally deterministic view of kinetic art as a material, mechanical, and formal practice, by comparison to the dematerialised, ephemeral and immaterial conceptual, interactive, and digital artworks.

Popper's argument is problematic because kinetic artists have often focused on both the material and immaterial effects of movement in their work. Popper instead assumes that there is an inevitable progression of artists moving from mechanical to digital media, thereby encouraging thinking about kinetic art as a practice that we've moved on from. This is because Popper draws a distinction between technological media where digital media is associated with immateriality and conceptual art, while restricting kinetic art only to formal mechanical motion.

However it is important to consider the role of movement and time in sculpture where artists attempt to explore both the material and immaterial effects of movement. In order to destabilise Popper's binary distinction in this chapter I consider the role of emergent movement and preacceleration as a means for considering the material effects of movement . Through Anthony McCall's *Line Describing a Cone* (1973) and Anish Kapoor's *Shooting into the Corner* (2008) I emphasise the role of movement as a process that arrives from a virtual incipience. This considers the use of movement in real-time as a productive, emergent, material and immaterial process.

While *Revolution—Forms that Turn* is a key example for articulating the way contemporary art history draws from avant-garde kinetic artists to inform audiences of the origins and developments of art since the industrial revolution, it is also important to consider works of this artistic generation outside of the setting of the Biennale. In Chap. 5 I address the early European avant-garde artworks by Moholy-Nagy and Gabo and, importantly, the philosophical approaches to movement and time by Bergson, as three key prominent influences of kineticism in art. Additionally, the works by Gabo and Moholy-Nagy are interpreted in ways that problematise Bergson's concept of duration and his resistance to mechanical representations of movement. For Bergson, the representation of movement in art not only provides a distortion of the subjective experience of duration, it also encourages these representations to be mistaken for actual time itself (1910). Gabo and Moholy-Nagy used kinesis as a tool for bringing attention to the present temporality through mechanical media. Moholy-Nagy and Gabo attempted to create new perceptions of temporality in real-time through movement, mediation, and distortion. This is particularly evident in their use of 'virtual volumes' (1965). These volumes are images of movement that are produced when an object moves rapidly enough to create a blurred formation of present and past movements.

Bergson, Moholy-Nagy, and Gabo's approaches to movement in art are also influential for contemporary artists interested in the phenomenological experience of movement and time in art. The emphasis on temporality in the virtual that I establish in Chap. 5 is also useful for my analysis of Olafur Eliasson's installation titled *Your Negotiable Panorama* in Chap. 6. Here, I draw a connection between Eliasson's approach to the virtual with Gabo and Moholy-Nagy's, to explore the material and immaterial effects of movement that emphasises time as a continual transformation in the present that also consists of folds of the past within it.

In this chapter I use *Your Negotiable Panorama* to consider movement in the present as a productive creation of actualisation from the virtual. This approach to time also resonates with a Bergsonian understanding of time, which, I argue, is remodernised by Eliasson. This is because they both privilege an experience of temporality that is experienced as a unified present that emerges from immanence. Consequently rather than using virtuality in art to move away from a materialised state, as Popper argues, there can also be alternative historical forms that continue to use movement as a tool for bringing a consciousness of time to the foreground. In regards to Eliasson, the virtual stems from Moholy-Nagy and Gabo, while he also uses movement to create slippages of indiscernibility between what is actual and virtual in the present.

In order to understand the roles, functions, and affects of kinetic sculpture that contribute to this consciousness of time, it is important to consider several features within the history of kinetic sculpture that work to marginalise the practice from contemporary interpretation. In this book I aim to reassess various ways in which contemporary and historical avant-garde kinetic sculpture not only features in the wider milieu of contemporary art, but is also a resource for audiences today to reflect on the conceptualisation of contemporaneity in art history.

As Pierre has recently described, there has been a clear critical absence of discussion around kinetic art and its interpretation in art history. This book therefore identifies a number of catalysts that have contributed to this absence and restricted kinetic art from contemporary art history and theory. Each of the chapters in this book focuses on a polemic that conflict with an understanding of kinesis as an expression of time in order to shed light on various roles and functions of kinetic art history within the wider milieu of contemporary art and its relationship to contemporaneity.

Understanding the roles and affects of contemporary and historical kinetic art in this way is significant for assessing the wider implications of contemporaneity today. Kinetic art is not necessarily the antiquated machinic avant-garde practice that those such as Popper frame it to be, and which is seen to have little relevance to contemporary society but is instead indicative of an ongoing relationship with time in art. The central argument of this book is that there needs to be a reconsideration of the relationships between kinetic art and time, and how these orchestrations of temporality are present in contemporary art and contribute to the multiple approaches to contemporaneity today. This reassessment of the contemporary and historical interpretations of kinesis in art also provides an opening for considering the multiple intersecting histories in the fields of technological, mechanical, electronic, and digital art. Rather than considering kinetic sculpture and installation as primarily an antecedent to contemporary media arts, my approach to the field points to a continued desire to use actual movement to express and explore the nature of contemporaneity through time.

References

Althusser L, Balibar E (1970) Reading capital (trans: Brewster B). NBL, London
Benthall J (1972) Science and technology in art today. Praeger Publishers, New York
Bergson H (1910) Time and Free Will: an essay on the immediate data of consciousness (trans: Pogson FL) S. Sonnenchein & co., London
Bergson H (1911) Matter and memory (trans: Paul NM and Palmer WS). Macmillan, New York
Best S (2012) Vibration, Vibração, Vibración: Latin American Kinetic art of the 1960s and '70s In the Powerhouse Collection.http://sydney.edu.au/museums/pdfs/Art_Gallery/VIBRATIONcatalogue.pdf. Accessed 30 Oct 2015
Bois YA (2000) Force fields: phases of the kinetic. Artforum 39(3):145–146
Borja-Villel M (2000) 'Preface', force fields: phases of the kinetic. Hayward Gallery, London
Bourriaud N (ed) (2009) Altermodern: tate triennial. Tate Publishers, London
Burnham J (1968a) Beyond modern sculpture: the effects of science and technology on the sculpture of this century. G. Braziller, New York
Burnham J (1968b) Systems esthetics. Artforum 7(1):30–35
Burnham J (1969) Real time systems. Artforum 8(1):49–55
Burnham J (1974) 'Systems Esthetics', Great western salt works: essays on the meaning of post-formalist art. G. Braziller, New York
Candy L, Edmonds E (2002) Context. In: Edmonds E (ed) Explorations in art and technology. Springer, London
Christov-Bakargiev C (2008) 'Revolutions forms that Turn: the Impulse to Revolt'. In: 2008 Sydney Biennale: Revolutions – Forms That Turn. Thames and Hudson, Sydney
Danchev A (ed) (2011) 100 artists' manifestos: from the futurists to the stuckists. Penguin, London
Deleuze G (1986) Cinema 1: the movement-image (trans: Tomlinson H, Habberjam B). Minneapolis University Press, Minneapolis
Deleuze G (1989) Cinema 2: the time-image (trans: Tomlinson H, Galeta R) University of Minnesota Press, Minneapolis
Derrida J (1984) No apocalypse, not now (Full speed ahead, seven missiles, seven missives). Diacritics 14(2):20–31
Desmond M (2008) Back to the future?—2008 Biennale of Sydney. Art Monthly Australia 2008 (213):5–9
Edmonds E (2011) 'Interactive Art' interacting: art, research and the creative practitioner. Libri Publishing, Faringdon

Eliasson O (2007) Take your time: a conversation. In: Grynsztejn M, Bal M (eds) Take your time, Olafur Eliasson. San Francisco Museum of Modern Art and Thames and Hudson, San Francisco
Foster H et al (1999) Questionnaire on "The Contemporary". October 130:3–124
Foucault M (1986) Of Other Spaces. Diacritics 16(1):22–27 (trans: Miskowiec J)
Gere C (2006) Art, time, technology. Berg Publishers, Oxford
Ghosts in the Machine. New museum. http://www.newmuseum.org/exhibitions/view/ghosts-in-the-machine. Accessed 30 Nov 2015
Grosz E (2004) Nick of time. Allen and Unwin, Sydney
Habermas J (1987) The philosophical discourse of modernity. (4th edn., trans: Lawrence F). The MIT Press Cambridge, Massachusetts
Jameson F (1991) Postmodernism: or, the cultural logic of late capitalism. Verso, London
Jameson F (2003) The end of temporality. Crit Inq 29(4):717–718
Jones C (2012) System symptoms. Artforum 51(1):113–114
Jim Allen, Len Lye and Hélio Oticica Form points of contact at the Govett-Brewster Gallery. Govett-Brewster art gallery. http://www.govettbrewster.com/TheGallery/MediaReleases/id/205/title/jim-allen-len-lye-and-helio-oiticica-form-points-of-contact-at-the-govettbrewster.aspx#expand. Accessed 13 July 2015
Kubler G (1962) The shape of time: remarks on the history of things. Yale University Press, New Haven
Lee P (2004) Chronophobia: on time in the art of the 1960s. MIT Press, Cambridge
Lefebvre H (1991) The production of space (trans: Nicholson-Smith D). Blackwell, Oxford
Lutticken S (2014) History in motion: time in the age of the moving image. Sternberg Press, New York
Magnaguagno G (2005) 'Foreword', moving parts: forms of the kinetic. Museum Tinguely, Graz
Manning E (2009) Relationscapes: movement, art, philosophy. MIT Press, Cambridge
Momin S (2008) 'Time Changes'. In: Huldisch H, Momin S (ed) Whitney Biennial 2008. exhibition catalogue, Whitney Museum of American Art, New York
Moholy-Nagy L (1965) Vision in motion. P. Theobold, Chicago
Nash M (2000) 'The Art of Movement'. In: Force fields: phases of the kinetic. pp. 313–316
Pierre A (2010) Instability: the visual/bodily perception of space in kinetic environments. In: Dezeuze A (ed) The 'do-it-yourself' artwork: participation From Fluxus to New Media. Manchester University Press, Manchester
Popper F (1975) Art, action and participation. Studio Vista, London
Popper F (1968) Origins and developments of kinetic art. Graphic Society, New York
Popper F (2007) From technological to virtual art. MIT Press, Cambridge
Riding A (2000) Arts abroad: retro or nostalgic, the work never stops moving. New York Times 2000. http://www.nytimes.com/2000/08/16/arts/arts-abroad-retro-or-nostalgic-the-work-never-stops-moving.html. Accessed 16 Nov 2016
Rush M (1999) New media in late 20th century art. Thames and Hudson, London
Schjeeldahl P (2012) Machine dreams: A Cabinet of Curiosities at the New Museum. The New Yorker, 6th August 2016. url:http://www.newyorker.com/arts/critics/artworld/2012/08/06/120806craw_artworld_schjeldahl. Accessed 29 Nov 2012
Shanken E (1998) The House that Jack Built. Leonardo Electronic Almanac 6(10). http://mitpress.mit.edu/e-journals/LEA/ARTICLES/jack.html. Accessed 10 Feb 2016
Shanken E (2004) Art in the information age: technology and conceptual art. In: Corris M (ed) Conceptual art: theory, myth and practice. Cambridge University Press, Cambridge
Shanken E (2012) Not just smoke and mirrors: a brief history of light and motion in art in relation to the work of random international. http://artexetra.files.wordpress.com/2009/02/smoke-and-mirrors-a4.pdf Accessed 26 Dec 2012
Shanken E, Ascott R (2003) Telematic embrace: visionary theories of art, technology and consciousness. University of California Press, Berkeley
Smith T (2008) Introduction: the contemporaneity question. In: Antinomies of art and culture: modernity, postmodernity and contemporaneity. Duke University Press, Durnham

Smith T (2009) What is contemporary art? Chicago University Press, Chicago
Smith R (2012) Technology advances, then art inquires: 'Ghosts in the Machine' at the new museum. New York Times. http://www.nytimes.com/2012/07/20/arts/design/ghosts-in-the-machines-at-the-new-museum.html?pagewanted=all&_r=0.
Soja E (1989) Postmodern geographies: the reassertion of space in critical social theory. Verso, London
Terranova CN (2014) Systems and automatisms: Jack Burnham, Stanley Cavell and the evolution of a neoliberal aesthetic. Leonardo 47(1):56–62. *Project MUSE*, muse.jhu.edu/article/537542.
Tinguely J (2011) Für Statik 1959. In: Danchev A (ed) 100 Artists' manifestos: from the futurists to the stuckists. Penguin, London, pp 336–337
Toffler A (1970) Future shock. Bodley Head, London
Virilio P (1984) The lost dimension. Semiotext(e), New York

Chapter 2
Revolutions—Forms That Turn

Abstract This chapter focuses on the sixteenth Biennale of Sydney titled *Revolutions – Forms that Turn* as a unique biennale that incorporated modern and contemporary kinetic artworks in order to reflect on time, revolution and movement in modern, postmodern and contemporary art. Consequently the biennale is useful for considering the changing conceptualisations of movement, temporality and contemporaneity. I argue that the curation of the biennale is emblematic of the peculiar relationship to time in contemporary art; a present that unfolds in time, while also accumulating and preserving the past within it, to which the role of actual movement in *Revolutions – Forms that Turn* were critical in forming.

Today biennales are the centre stage for contemporary art in the Artworld. Not only do they showcase and discuss the nature of our contemporaneity, but they also differ from the modern world fairs and biennales, because they explicitly project multiple fractured histories and identities. The desire to overturn previous political and theoretical structures is central to the discussions around contemporary biennales. For instance, when asked 'what makes a biennial?' world-renowned curator Rosa Martinez answered that 'The idea biennial is a profoundly political and spiritual event. It contemplates the present with the desire to transform it', and is indicative of a larger social, political and economic flows within contemporary society (Stallabrass 2003, p. 34). We might even say that biennales are perceived as trendsetters, or predictors of intercultural flows that focus on the political nature of art in a global setting.

Considering this perception, the sixteenth Biennale of Sydney *Revolutions—Forms that Turn* was an unusual biennale at the time. Rather than focusing solely on contemporary artists the biennale presented modern, postmodern and contemporary artists alongside one another in order to reflect on the themes: time, revolution, and formal and political movement in art in the past century. It is no surprise then that the biennale was criticised by some for reflecting more on art history, than contemporary art (Desmond 2008, p. 5).

Despite this criticism, another unique aspect to the biennale was the incorporation of kinetic sculpture throughout the exhibition spaces. Viewers were

© Springer Nature Singapore Pte Ltd. 2017

C. Chau, *Movement, Time, Technology, and Art*, Springer Series on Cultural Computing, DOI 10.1007/978-981-10-4705-3_2

confronted with mechanical sculptures that stuttered in loops, rotating spikes, swaying mobiles, and swung fans across a ceiling. The biennale also coincided with a number of exhibitions in the last few years that focused on actual movement in art in western art history elsewhere in the world. Many of these exhibitions celebrated the presence and influence of kinetic sculpture in art history and contemporary society. For example, *Moving Parts: Forms of the Kinetic* (2004) at the Graz Kunsthaus focused on mechanical movement in art that has 'come to appear as an everyday normality to us' (Magnaguagno 2005, p. 9); at the New Museum in New York *Ghosts in the Machine* presented kinetic sculpture as a 'prehistory to the digital age'; *Force Fields, Phases of the Kinetic* (2000) at the Hayward Gallery was 'an investigation of movement in art' to remind audiences of the presence of mechanical movement in sculpture prior to the digital age (Nash 2000, p. 313).

As mentioned in the introduction of this book, kinetic sculpture is often remembered as an antiquated exploration of movement in sculpture, and conjures associations of being a 'cabinet of curiosities' dug up from a forgotten age. Therefore Christov-Bakargiev's decision to highlight the importance of kineticism in western art history during the Sydney biennale invites questions around how and why kinetic sculpture says something about contemporary art and society, and might be re-remembered by audiences. In the biennale kinetic sculpture was not regarded as an obsolete, mechanical and unrequited practice, as others have (Weibel 2007, pp. 21–41), but instead was considered as integral to our regard for form, movement, and time in the mechanical and digital ages.

What might be one of the reasons for Christov-Bakargiev's curatorial decision? Rather than solely focusing on new and emerging artists Christov-Bakargiev presented audiences with reflections of the past in order to better understand the present. In effect *Revolutions—Forms that Turn* sensationalised elements of modern avant-garde art within a contemporary context and reimagined kinetic sculpture as a signifier for reflecting on time, movement, form and revolution. Consequently the biennale now helps us to think about how kinetic sculpture has been framed by art history in the past, and how these analyses have changes over time.

Revolutions—Forms that Turn indicates some of the changing roles, functions and effects of exhibiting modern kinetic art in contemporary exhibitions. That is, rather than an obscure and obsolete avant-garde tendency, the curation of this biennale is useful for considering contemporary and historical kinetic art as a resource for contemplating the changing conceptualisation of movement, temporality and contemporaneity. The biennale is also valuable for tackling the assumption that kinetic sculpture is a key precursor to contemporary digital art. This idea is most prominently suggested by Frank Popper on multiple occasions including most strongly in *From Technological to Virtual Art* and *Origins and Developments in Kinetic Art*.

Revolutions—Forms that Turn is one of many large-scale international exhibitions that have drawn from avant-garde kinetic art in order to reflect on contemporary art theory and history in relation to perceptions of movement and temporality. Reflecting on the relevance of avant-garde kinetic sculpture in contemporary society, this chapter uses the biennale as an example for considering various facets of kinetic art history from a contemporary perspective that conflict with past interpretations of the artistic practice. If, as I argue in this book, kinetic art helps us to think about the changing conceptions of temporality, over time, *Revolutions—Forms that Turn* is also valuable for opening new questions concerned with the role and effects of historical and contemporary movement art.

2.1 Revolving, Rotating, Mirroring

The title of the biennale, *Revolutions—Forms that Turn* resonates with what Marshall Berman has described in *All That is Solid Melts into Air: The Experience of Modernity* that there is a natural affinity between modernity and revolution, because modernity depends on technological and political progress conquering its ancestors. Berman reminds us that the term 'modern' depends on conquering the old, and associates all change with human advancement. In Berman's words: 'what is modern is newer than what it succeeds: what is revolutionary is more advanced than what it overthrows—"tradition" in one case, "reaction" in the other' [47].

Revolutions—Forms that Turn echoed a similar regard for tradition, reaction, and revolution in modernity. The exhibition drew from historical canons of art in order to engage with the elements of contemporary art and society in a self-reflexive manner. Avant-garde artists were presented alongside contemporary counterparts to identify that change, revolution, innovation, and progression have been long standing central themes in society since modernity. The artistic director, Christov-Bakargiev presented both historical and contemporary works as part of a continued social impulse of revolt. As Christov-Bakargiev explained:

> **What is the theme of the Biennale?** The impulse to revolt. Revolving, rotating, mirroring, repeating, reversing, turning upside down or inside out, changing perspectives. Such literal and formal devices are charted for their broader aesthetic, psychological, radical and political perspectives. This project explores the relationship and distance between 'revolutionary art' and 'art for the revolution', the space between form, on the one hand, and the role of art in society on the other (2008, p. 30).

Christov-Bakargiev showed a clear intention for the biennale to be an event for reflection, critique, and new interpretation of the contemporary world. The theme and statement nurture reflections and inquisitions into how the meaning and expression of political revolution has been a key characteristic in art in the nineteenth and twentieth centuries. Consequently *Revolutions—Forms that Turn* also focused on broad themes that have often been associated with modernism such as temporality, newness, mechanical reproduction, revolution and the avant-garde.

While a number of art historians, critics, and scholars have argued that contemporary art and society has moved beyond modern and postmodern approaches, others such as art historian Terry Smith have taken a less linear approach. In *What is Contemporary Art?* Smith identified that contemporary art has a number of coalescing trends, one of which focused on continuing issues and debates that have persisted since modernity. For Smith this is a tendency that tries to 'cleave new art to the old modernist impulses and imperatives, and renovate them' (2009, p. 7). Direct examples of these trends at play includes art institutions such as the Museum of Modern Art (MoMA), which explicitly focuses on presenting modern art to contemporary audiences as a way to informing audiences on the heritage of contemporary art and society.

Similarly *Revolutions—Forms that Turn* emphasised the importance of the works by modern artists into a contemporary exhibition space to remind audiences of an art historical lineage that moves from modern to contemporary art. In effect, one might argue, as Christov-Bakargiev did, that the exhibition was simultaneously modern and contemporary. Such a curatorial decision might not seem so groundbreaking or surprising, considering other institutions like MoMA, Dia Art Foundation, the Tate Gallery also function in a similar way. However, unlike museums and major art institutions, biennales are not traditionally known for educating audiences about art history. Rather, as mentioned earlier, biennales today are geared towards to examining and debating issues laden in present art and society today.

Therefore the focus on the past transforming the present is what sets *Revolutions—Forms that* Turn apart from other biennales in recent years. The number of modern avant-garde artworks that were featured dominated over the presence of contemporary artists. Among the modern avant-garde works there was a concentration on early modern Constructivist, Dadaist, and Futurist artists, which included key art historical works from the early twentieth century that represented the machine aesthetic within modern art and modern mechanical movement . These included the 'Futurist Manifesto' by Filippo Tommaso, Marcel Duchamp's *Bicycle Wheel* (1920); his collaboration with Man Ray, *Rotary Glass Plates, Precision Optics* (1920), mobiles by Russian Constructivist Alexander Rodchenko such as *Hanging Spatial Construction* Nos. 9–13 (1921-21), and Alexander Calder's hanging kinetic mobiles *Hanging Spider* (1940) and *Roxbury Flurry* (1946). Already recognised as experimenters of social and political change through their art, these artists were presented in the biennale as reminders of the ancestral lines of avant-garde works that constitute contemporary art history (Engberg).

Also at the biennale were many artworks that were created in the decades shortly after WWII and responded to the technological and social issues at the time. These earlier artists were joined with central avant-garde figures from the 1960s to the 1980s, and were projected by the biennale as works through which to consider the previous approached to revolution, change, repetition and re-enactment in relation to the other artists (Christov-Bakargiev 2008, pp. 30–31). This includes John Cage's durational work *4'33"* (1952/2004), Guy Debord's film *Le Société du Spectacle* (1973), documentary material of Chris Burden's performance *Shoot* (1971), Carolee Schneeman's *Meat Joy* (1964), Valie Export's *Touch Cinema* (1968), and Gordon Matta-Clark's film, *Program Eight: Office Baroque* (1977/2005). These works contributed to a landscape of late-modern art during a period when the role of the avant-garde becoming an artistic tradition of its own, while also attempting to challenge traditional modes of representation in art.

Within the exhibition historical and contemporary kinetic sculpture played an integral role in implementing discussions about the modes and methods of how art history is framed, changing perceptions of time and technological aesthetics. Many reviews of the show also specifically commented on the presence of kinetic sculpture throughout the event (Clement). This is another reason why there was a stronger presence of historical works than those by contemporary artists in the biennale.

In *Revolutions—Forms that Turn* historical and contemporary kinetic sculpture played an integral role in implementing discussions about the modes and methods of art history, changing perceptions of time and technological aesthetics. This curatorial decision reinserts kinetic sculpture into a larger framework of Western art history and disassociates the practice as an unrequited and forgotten phase of modern art. Consequently the exhibition itself stands as a reconstruction of art history, within which kinesis is used as a tool for social and political revolution, as well as the revolution of aesthetic form, and this has continued in art throughout capitalist society.

2.2 Cum-Tempore (with Time)

The incorporation of kinetic sculpture in the biennale helped to convey the central curatorial themes that Christov-Bakargiev wanted to communicate and use to spark discussion: movement, revolution and repetition. Many of the sculptures were formal expressions of these themes, but laden in them were also tones of expression around political movement, revolution and the idea that history repeats itself. These themes encouraged questions such as 'what does revolution look like today?', 'what has changed in society over time?' and 'will we make the same mistakes as our ancestors?'. Reflecting on history, political revolution and movement in the machine age meant that time, reflection, memory and temporality were central to the biennale, and audiences were encouraged to think about the changing attitudes of time, over time.

Each culture internalises its own peculiar concept and approach to time, and it has been suggested that western society has done so according to each technological age. For instance, modernity has often been characterised by the movement of mechanical acceleration, and time that moves in a linear fashion with the future moving through the present and into the past; While postmodernity enabled people to explore the instantaneity of digital technologies, and multi-linear time codes.

Similar trends in western art history have also prevailed, especially when considering artists who use mechanical and digital machines in their practices. When thinking about contemporary conceptualisations of time in art, Terry Smith provides a useful perspective that deals with the multiplicity of technologies and cultural flows that prevail today. In *What is Contemporary Art?* Smith describes the current approach to time as:

> characterised more by the insistent presentness of multiple, often incompatible temporalities accompanied by the failure of all candidates that seek to provide *the* overriding temporal framework – be it modern, historical, spiritual, evolutionary, geological, scientific, globalizing, planetary…Everything about time these days – and therefore about place, subjectivity, and sociality – is at once intensely *here*, is slipping, or has become artifactual (p. 196).

Smith paints a complex relationship to time which is elusive, 'slipping' away, and containing a multiplicity of time codes (p. 196), which is indicative of the contemporary information age where we are privy to many kinds of competing and conflicting knowledge systems and notions of time.

Smith also gives opportunity to thinking about a multiplicity of time codes that reside in the present. It is precisely this multiplicity that characterises our relation to being in time in contemporary society. Similarly, rather than thinking about time in a linear fashion where the future draws into the present and become the past, the biennale encouraged audiences to consider the present temporality as a reserve of multiple time codes within it. As such contemporary art is not necessarily associated with newness, progression, being present, or up to date. Christov-Bakargiev is interested in the present being more than a singular 'now', but which gathers folds of the past within it. As Christov-Bakargiev explains: 'everything that exists in the world is of my time, whether it is an old 1950s Bakelite telephone, or an artwork made two years ago or today' (p. 33): we gather memories of the past into the present in order to inform, reflect and understand identity in contemporary society. Consequently this idea lends us to think that contemporary society is defined by what it remembers, because the past is continually reframed in time by contemporary perceptions and interpretations: the present unfolds in time, and does so by drawing on the past within it.

Christov-Bakargiev's interest in how we might fold the past within the present resonates with Michel Serres' approach to contemporaneity. Serres uses the example of a late-model car as an accumulation of inventions from the past as an analogy for defining 'contemporary' (p. 45). The wheel, motor and every other component were all separately invented in different eras, and these have come together with the invention of the car. In effect, the car is an accumulation of past milestones. While the contemporary contains the past, the only true indication that the car has been made in contemporary time is the assemblage of any additional inventions, its design, advertising, consumption and ideology.

Unlike Smith and Christov-Bakargiev, Serres does not privilege the contemporary but rather eradicates distinctions between the historical and the contemporary. He projects an absolute 'indifferen[ce] to temporal distances', where everything is at once contemporary *and* historical (Serres 1995, p. 45). While Christov-Bakargiev might not reach as far as Serres' collapse of temporal distinctions, both regard the present as a force that folds the historical within it. The difference between Serres with Smith and Christov-Bakargiev is that the latter still focus on present relationships with time, and therefore privilege the contemporary. This implies that constructions of art history today, according to Christov-Bakargiev, are constituted as contemporary, but not necessarily vice versa.

2.3 Avant-Garde Kinetic Artworks at the Biennale

Key players in the history of kinetic art were a key feature in executing the ideas of the biennale. Many of these artists were concerned with the movement of mechanical sculpture in order to think about the futility of industrial labour, and explore the performative possibilities of sculpture. These included Jean Tinguely's *Bascule no. 1: Sisyphus* (1965) and *Méta Malevich* (1954) and remnants of *Homage to New York* (1960), the latter of which were salvaged after a performance where Tinguely assembled scrap metal and goods including a piano, bicycle wheel, canisters of paint installed in the sculpture garden of MoMA before it played out its destruction: an explosion of waste to critique excess, consumption and commodification in capitalist societies. While *Homage to New York* might be regarded as a kinetic performance rather than a kinetic sculpture, the artwork resonates with Tinguely's *Bascule no. 1: Sisyphus*, and *Méta Malevich*—both mechanical sculptures doing what machines should not do in a utilitarian society: move without producing.

Len Lye's *Storm King* (1964) and *Ribbon Snake* (1965/2008 a reconstruction was created specifically for the biennale), continued discussions around mechanical reproduction. Both kinetic sculptures repeatedly move in rotations at regular intervals but also create irregular shapes, and sounds. Like Tinguely, Lye's machines were nonsensical, unproductive (in the utilitarian sense) and repetitive forces of movement. Lye used kineticism to explore the effects of movement as an organic force, even which mechanical materials could obtain.

David Medalla's *Cloud Canyons* (1967/1985) include a jet of soapy foam building up through a series of cylinders and projecting outwards through each tube. Despite being a mechanical sculpture the form of the foam is repetitive, yet unpredictable; constantly changing and yet unrepeatable in order to rupture the utilitarian association with mechanical works.

These artists have frequently been commended for their experimental techniques with mechanical form as well as social and political ideology. The decision to disperse contemporary artworks among a predominantly historical exhibition alludes to contemporary art as a continuation of modernism, or what Terry Smith would term as 'remodernism'. *Revolutions—Forms that Turn* reifies that the contemporary artworks signify a cyclical return to artistic experimentation, political antagonism, and dissolving traditional modes of representation. Additionally, this curatorial composition infers that form and aesthetic move in cycles; that post-modernity is historically read as a part of the greater cycles of art, and modernity as an ideology that is ongoing, even endless.

Among the contemporary artists that featured in the biennale, mechanical movement was also a regular feature. This included Olafur Eliasson's *Light Ventilator Mobile* (2002) and Rebecca Horn's *Cutting Through the Past*, both of which I will focus on in the following section; Michael Snow's *De La* (1969–1972), a rotating sculpture fitted with surveillance cameras, monitors, and controls that stalk viewers and project their movements on a screen without delay. Also exhibited was a reconstruction of Vladimir Tatlin's *Monument to Third International* (1919) entitled, *White Man Got No Dreaming* (2008) by Michael Rakowitz constructed with scrap material from what was the Pemulwuy Aboriginal Housing redevelopment plan for Redfern in Sydney. The works aptly echoed the biennale's themes around revolution, rupturing tradition, and resistance.

Exhibited alongside Calder's black-and-white hanging mobiles was Eliasson's *Light Ventilator Mobile*: a hanging, rotating mobile that balances the weight of a domestic fan on one end of a rod, and hanging down with its electrical chord on the other end is a spotlight shining onto the walls of the space. The motorised mobile rotates rapidly, while the dangling fan swings unpredictably at eye level, and the shine of the circling light attached at the opposite end of the mobile catches and briefly blinds the viewers. This work is dizzying, and tests the threshold between chaos and predictable circular action. Rather than conjuring political engagement like Snow and Rakowitz, *Light Ventilator Mobile* provides a more formal exploration of movement, chaos and predictability. There is a sense of displaced movement between the erratic motion of the fan and the comparatively smooth yet unpredictable glide of the light beam because a single horizontal rod connects them. Additionally, the movement of the viewer is restricted by the performance of the mobile. The rotation of the mobile casts a territorial circumference within the exhibition space, which viewers are not usually inclined to enter because of the peril of the swinging fan. The speed of the light beam extends the space occupied by the sculpture, casting outwards, rather than inwards, encouraging viewers to further step further back, and watch the light tracking the walls around it.

Eliasson is well known as a contemporary artist concerned with the phenomenological relations between the viewer and object and the act of perceiving as an uneven and, at times, waning and swelling subjective experience where one is 'seeing yourself seeing' (Lee 2007, 'Your Light and Space', pp. 33–35). When considering temporality Eliasson describes history as 'not external and objectified in a situation but… inside the spectator' (Eliasson in Morgan 16), as a constant reconstruction that is borne by each viewer and their experience. After the biennale, Eliasson also produced a number of works that expressed his curiosity with orchestrating temporality through kinetic lumia, sculpture, and installation, and exhibited them in the Museum of Contemporary Art in Sydney, a main venue for the Sydney Biennale. His exhibition *Take Your Time* focused on the subjective perceptions of and within the present temporality. He suggested that '[t]he sense of time that I work with is the idea of a "now"….There is only a "now"…our belief in time is just a construct'(quoted in Alle 2000, p. 30). Like many artists who have used actual movement to express time before him, Eliasson expresses sensitivity towards the effect of the temporal present kinetic movement on his audiences. He aims for his works to let viewers ‘see themselves seeing’ this ‘now’ (Lee 2007, ‘Your Light and Space, p. 35), similarly to early avant-garde artists who used movement to draw attention to perceptions of the present.

Rebecca Horn’s installation in the biennale was as perilous as Eliasson’s swinging fan. Her *Cutting Through the Past* (1993–1994) is constructed with five doors standing in a circle on a platform with the edges of their frames pointing inwards towards a horizontal spike that rotates in the centre. At each rotation the spike scratches into the edge of each doorway as it passes, with what has been described as an ‘incisive and cruel gesture, rich with erotic implications’ (Engberg). Horn’s motion produces conflicting effects because each rotation is performed in its predictable manner, and yet as the spike approaches each doorway, a moment of piercing tension is orchestrated. *Cutting Through the Past* functions similarly to Eliasson’s *Light Ventilator Mobile* as both artists manipulate intensity and the accumulation of tension and intensity even when mechanical movement is constructed to move in a repetitious manner.

These contemporary gestures play to Christov-Bakargiev’s cyclical approach to time, which is in constant progression, and the effects of the modern machine aesthetic continue to be effective media for expressing time rhythms. Horn and Eliasson perform mechanised activities that range from the carefully measured, to entropic and unpredictable, and explore tensions laden within the present—time moves in obvious circles, and yet moments of tension, energy and suspense can arise within it. The curatorial decision to exhibit these works in the same biennale alongside the avant-garde works by Lye, Alexander Calder, Tinguely, and Medalla presents an assumption that Eliasson and Horn continue this artistic investigation. Although to varying degrees the reliance on the movement of form to build tension continues a desire to explore the actual and virtual effects of kinesis that was established by modern artists. These earlier artists from the 1960s (and in Calder’s case, the 1930s), each explore the capacity for unpredictable movement and unforeseen forms through repetitive mechanical actions.

As an art historical exhibition, Juliana Engberg claims that the biennale constructed political revolution as a recurring and yet weakening element of present society. As she summarises, ‘[a]t the beginning of the twentieth century, represented by Rodchenko, art is launching into space, daring us to embrace the dawn of speed and velocity. By the end of the century, represented by Cattelan’s forlorn hoisted horse, all the puff has gone out of the revolutionary enterprise’(Engberg). Consequently the biennale drew together an art historical construction that reified the spirit of revolution with the modern avant-garde, and undermined contemporary art as a site for political engagement (Desmond 2008, p. 5). Therefore, to move forward in time requires a reconstruction of modern art that resensationalises it within contemporary contexts.

This interpretation also positions *Revolutions—Forms that Turn* as a critical inquiry into the dialects between modernity and postmodernity, and an attempt to collapse the distinctions between these canons. Consequently, kinetic sculpture and installation are significant players in the reconstruction of contemporary art historical interpretation. For instance, Horn and Eliasson’s work are situated as extensions of the modern machine aesthetic presented by early European avant-garde artists. Furthermore, as Christov-Bakargiev has argued, these early modern works of art are contemporary precisely because they are perpetuated by artists, institutions, and exhibitions as forerunners of the present sense of contemporaneity.

In this respect *Revolutions—Forms that Turn* is a contemporary exhibition that returns to and perpetuates avant-garde tropes to contemporary audiences. The modern aesthetic is for this reason part of the contemporary because Christov-Bakargiev's decision to use movement as a tool for reevaluating contemporary art history is what Smith would describe as a 'remodernist' action (Smith 2009, pp. 5–7). However, more importantly for the focus of this book, an effect of this remodernist perspective is that a reevaluation of the roles and effects of kinetic art in relation to understandings of movement and time are encouraged. *Revolutions—Forms that Turn* is indicative of a renewed art historical evaluation of kineticisim in practice. The biennale aligned modern avant-garde mechanical kinetic works alongside contemporary mechanical installation, thereby strengthening the association between kinesis as a continued tool for contemporary artists. Rather than bringing together contemporary artworks that represented a digital contemporary avant-garde, Christov-Bakargiev made a conscious decision to present to the international artistic community a disregard for digital determinism, and presented modern industrial kinesis as an informant to contemporary art historical canons and characteristics of contemporaneity today. This encourages new relationships between avant-garde kineticism and the contemporary to be formed. Considering Christov-Bakargiev's approach to contemporaneity, *Revolutions—Forms that Turn* is presented as a contemporary art historical view that does not necessarily frame technological art as a move from mechanical towards virtual art, but builds as a multiplicity of time codes and the machine aesthetics within it.

This perspective results in an opportunity for interpreting kinetic art history that resists the assumption that mechanical kinetic sculpture is a rudimentary form of digital art in motion. Another consequence of the arrangement of *Revolutions—Forms that Turn* is for the dichotomy between the mechanical and the digital to be destabilised. Considering Christov-Bakargiev's approach to contemporary art history, the artists in the exhibition were arranged as pioneers of contemporary art and were themselves to be regarded as contemporary. Kinesis was therefore used in the biennale as a tool for demonstrating that the modern avant-garde is at once considered historical and contemporary in contemporary art. This challenges the idea of digital art as the 'more refined' technological art to modern mechanical media, because as I have highlighted, the biennale indicates a construction of art history today that collapses temporal distinctions between the historical and the contemporary and continually reconstructs by drawing from the recent past.

While Horn and Eliasson's works are engineered with motors that are not experimental by present-day standards, technological media is used as a means of achieving their formal and conceptual intentions. This is unlike the kinetic artists from the 1920s or 1960s in the biennale who were concerned with building or critiquing a machine aesthetic in their artworks. Christov-Bakargiev's perspective of contemporary art, one that is *with-time*, indicates, as I will argue throughout this book, that there is still a strong desire to express and consume modern and contemporary kinetic sculpture because of its association with temporality. The contemporary works such as those by Eliasson and Horn use movement and time to assert identity through a process of entropy that attempts to locate and dislocate viewers through the use of motion (Smith 2009, p. 7). Both artists depend on entropy, control, movement and duration as ways to draw attention, and affect their audiences in spectacles of movement similar to the early European avant-garde kinetic works they were presented alongside with at the exhibition. They are playful the limits of human perception through experimentation of form, and depend on shocking their audiences with suddenness to create new sensations.

Christov-Bakargiev's direction is also useful for unpacking the peculiar expressions of temporality in modern art to contemporary audiences. *Revolutions—Forms that Turn* returns to the space-time conceptions that were popular at the time of early avant-garde, while also encouraging discussions about the use of movement and expression of time through contemporary kinetic sculpture. Unlike Bourriaud's proposition of a new modernity, which he called 'altermodern' (2009), she is concerned with the contemporary artists that signify one of many returns to modern avant-garde tendencies such mechanical aesthetics, and the use of shock, chaos, and repetition. This return is also considered as a greater condition of the cycles of form and aesthetic in art history, and is demonstrated by kineticism, specifically through works such as those by Eliasson and Horn next to Tinguely, Medalla, Lye, and Duchamp. This presents actual movement in art as a key concern for modern art history, and just as importantly, as a site for understanding of contemporary art. Christov-Bakargiev's approach compliments Smith's emphasis of art institutions that contemporise modern art, as well as contemporary artists perpetuating modern aesthetics.

Terry Smith's description of contemporary art today is useful for understanding some of the mechanisms behind the return to modern kinetic art in these three exhibitions. Smith's claim that a remodernist tendency is one of many emerging trends within contemporary art today is one explanation behind the desire to reassess art history, and reframe the influence kinetic sculpture has had on the artworld today. The biennale can therefore be considered as an engagement with contemporary audiences by resensationalising modern tropes of art such as revolution, shock, temporality and the machine aesthetic.

The relationship between time and mechanical movement in *Revolutions—Forms that Turn* in contemporary art history is valuable for discussing further the changing interpretations of modern kinetic sculpture, and confront the art historical assumption that considers modern kinesis as a precursor to digital art. Christov-Bakargiev's provides room for considering alternative historical narratives that run parallel to dominant narratives linked to digital art. *Revolutions – Forms that Turn* is a unique example of a contemporary art institution forming and reconstructing art historical lineages, not only because of the scale and breadth of the exhibition, but also because, as I have argued, it stands as an explicit divergence from traditional curatorial plans of biennales. Rather than focusing on emerging and ever-changing contemporary artistic tendencies, as most biennales do, the sixteenth Biennale of Sydney encourage reflections on contemporising modernity, as well as drawing divergences and references between contemporary art and art history.

2.4 Conclusion: Contemporary Exhibition of Modern Kinesis

Christov-Bakargiev's curatorial direction at *Revolutions—Forms that Turn* has projected a nature of contemporaneity and contemporary art history that collapses temporal distinctions among modern, postmodern, and contemporary periods in art. The biennale is an articulation of the present as a process that unfolds in time, while also accumulating and preserving the past within it. This arrangement encourages considerations about movement , particularly a reflection on the cyclical, linear, or heterogeneous ways art history has explored the tropes of revolution, form, movement, and time. By exhibiting a higher concentration of modern avant-garde than contemporary art, construction of contemporary art history that depends on the above-mentioned modern tropes is emphasised. These approaches to temporality and contemporary art history also open new ways to think about the artists and their works that feature in the biennale, in relation to contemporary art and the social and political milieu. An effect of this spectacle of modern kinetic art in contemporary art history is that it re-situates kineticism as a driver of contemporaneity today that resensationalises modernism. This perspective suggests that actual movement in art can be used as a resource for unpacking aspects of contemporary art and society today, particularly in regards to a number of approaches, expressions, and conceptualisations of time today. The biennale enframes an historical lineage between modern kinetic sculpture and contemporary installation that breaks away from associations of kinetic as a purely modern machinic tendency in art, but is also valuable to unraveling aspects of contemporaneity today. Therefore, a reconsideration of how artists have used movement to express and engage with time in the past is valuable for understanding contemporaneity in contemporary art.

Revolutions—Forms that Turn is a significant exhibition that brought to the fore discussions about the roles and effects of kinetic art for contemporary audiences. This is particularly in relation to sculptural movement as a way to engage audiences and influence their understanding of what it is to be with time in art. The next chapter addresses this connection in more detail. It presents the 1960s as a significant period because kinetic artists in the decade of the 'kinetic kraze' were devoted to mediating time with technology in different ways. The technological expansion after WW2, and the domestication of communications and computing technologies, gave rise to new polemics and perceptions of movement and time, and this is significant for understanding how and why the term 'kinetic' has been framed by contemporary art history.

References

Alle I (2000) Putting the viewer first: Olafur Eliasson, *Sculpture* 20(8)
Bourriaud N (2009) Altermodern: tate Triennial. Tate Publishing, London
Christov-Bakargiev C (2008) Revolutions forms that turn: the impulse to revolt. In: 2008 Sydney Biennale: revolutions—forms that turn. Thames and Hudson, Sydney
Clement T (August 2008) Mind the flying spike. Sydney morning Herald. http://www.smh.com.au/news/biennale/mind-the-flying-spike/2008/07/31/1217097404488.html. Accessed 14 Nov 2016
Desmond M (2008) Back to the future? – 2008 Biennale of Sydney. Art Monthly Australia 2008 (213):5–9
Engberg J (2008) Children of the Revolutions [The Sixteenth Biennale of Sydney]. Monthly. http://www.themonthly.com.au/art-juliana-engberg-children-revolutions-sixteenth-biennale-sydney-1137. Accessed 2 Nov 2016
Lee P (2007) Your light and space. In: Grynsztein M (ed) Take your time: Olafur Eliasson. Thames and Hudson, San Francisco
Magnaguagno G (2005) 'Foreword', moving parts: forms of the kinetic. Museum Tinguely, Graz, pp 6–11
Nash M (2000) 'The Art of Movement', force fields: phases of the kinetic. Hayward Gallery, London, pp 313–316
Serres M, Latour B (1995) Michel Serres with Bruno Latour; conversations on science, culture and time (trans: Lapidus R). The University of Michigan Press, Ann Arbor
Smith S (2009) Contemporaneity in the history of art: a Clark workshop, summaries of papers and notes on discussions. Contemporaneity 1(2011). http://contemporaneity.pitt.edu/ojs/index.php/contemporaneity/article/view/32/3. Accessed 18 Mar 2016
Smith T (2009b) What is contemporary art? University of Chicago Press, Chicago
Stallabrass J (2003) The aesthetics of net. art. Qui Parle 14(1):61
Weibel P (2007) It is forbidden not to touch: some remarks on the (Forgotten Parts of The) history of interactivity and virtuality. In: Grau O (ed) MediaArtHistories. MIT Press, Cambridge

Chapter 3
Directions in Kinetic Art: Expanding Spectres of Time

Closely related to the space problem in sculpture is the problem of Time…the latter still remains unsolved, being complex and obstructed by many obstacles
(Gabo, *Gabo on Gabo* 112)
The moderns were obsessed with the secrets of time
(Jameson 1984 *End of Temporality* 697)

Abstract This chapter frames the 1960s as a decade that experienced a turbulent relationship to time in art due to a wider technological transition between mechanical and digital machines. Consequently the 1960s is considered as period of temporal turbulence in contemporary art history, where artists used kinesis to reflect on this technological transition with utopianism, uncertainty, and resistance. The chapter highlights the tensions between movement, and technology via three key examples; Michael Fried's seminal essay 'Art and Objecthood'; the exhibition curated by Peter Selz titled *Directions in Kinetic Art* in 1965 at the Berkeley Museum; and two works by Jean Tinguely *Homage to New York* (1960) and *A Study for an End of the World* (1961/1962). These three examples illustrate the tensions in art theory and criticism regarding the role or time and technology in art at the time and argues that artists working with movement in their work were primarily interested in exploring new ways to conceive and experience time through technological media in art, as well as the temporal nuances and time codes that each machine age fosters.

The 1960s is remembered as a key decade in the twentieth century that grappled with technological, social, and political transition within the 'turbulent era' of modernity (Selz 1985, pp. x–xvi*)*, which experienced a 'peculiar form of acceleration' (Kocelleck in Lee *Chronophobia* xii) . This time of peculiar acceleration has also influenced the nuances of contemporary art, theory, and history today. As Pamela Lee has identified, 'what might seem a sixties problem is embedded in the web of our own present … its implications get played back—like a tape-loop—between our contemporary moment and that of the recent past' (p. xvi). For Lee, art history today specifically draws from this recent past in order to frame the foundations of contemporary artworld. The contemporary is therefore characterised in

© Springer Nature Singapore Pte Ltd. 2017

C. Chau, *Movement, Time, Technology, and Art*, Springer Series on Cultural Computing, DOI 10.1007/978-981-10-4705-3_3

part, by the return and reflection of art, theory, and exhibition in the decade of the 1960s.

The 1960s is also remembered as the decade where kinetic art had a brief burst of popularity, because it experimented with new technologies in art while also appealing to mainstream audiences. Kinetic sculpture was a popular during this time that (despite its popularity in Europe previously) came and dissipated quickly due to the growing antiquity of modern industrial media in a new technological age (Bois 2000, p. 145).

This chapter focuses on the 1960s as a period that experienced a turbulent relationship to time in art due to a wider social and technological transition that was experienced in society. Consequently it is framed as decade of temporal turbulence in contemporary art history, where artists used kinesis to reflect on this technological transition with utopianism, uncertainty and resistance. While art historian Charlie Gere has argued that the artistic responses to the increasingly accelerated society throughout the nineteenth century is a significant element of modern art (Gere 2006, p. 2), I argue that the 1960s is an especially important decade of debate because it was the nexus between two technological societies: the mechanical and digital age.

This chapter will focus on the increasing use of new technological media in art that was often used to discuss the changing conceptions of temporality in society, which brought new perceptions of the role of time and technology for an increasingly ontologically unstable definition of a work of art (Palmer 2008, pp. 156–184). Secondly, Michael Fried's seminal essay, 'Art and Objecthood' is a quintessential example of the turbulent relationship to time in art that was expressed during this period. Fried's perspective presents the hesitancy towards open durational artworks, because he considers time in art as a threat to modern formalist understandings of a work of art.

Following 'Art and Objecthood', I address the exhibition *Directions in Kinetic Art* (1965) held at the Berkeley Museum of Art as a key event that discussed kinetic art as a practice that had potential to create new perceptions of time in art (Selz 1985, pp. 275–285). Although the exhibition was not the earliest to feature moving sculpture, it was one of the first to discuss and debate the aesthetic of movement through technology in art in the 1960s (Selz *A Turbulent Era*, pp. 275–285).

For instance, the curator Peter Selz, and the artists in the exhibition, discussed movement as an artistic tool to explore the differences between the sensations of mechanised time and the subjective experience of time. What is significant about the discussions that emerged through the exhibition and its corresponding symposium is the frequent return to the relationships between time, movement and technology. From the documentation of the exhibition, it is clear that the term 'kinesis' was used by Selz, with the intention of cultivating sensitivity towards movement as an expression of time in an era of social and technological turbulence (Selz 1985, pp. 275–285).

Finally, while Jean Tinguely's early kinetic performances *Homage to New York* (1960) and *A Study for an End of the World* (1961/1962) are often addressed as critiques of capitalist production and consumption, this chapter will highlight these auto-destructive works as performances that orchestrate specific experiences of time. In both works Tinguely brings an awareness of the passing present to the fore, and problematises Fried's argument against durational works of art. These mechanical performances are also useful for highlighting how each technological age helps to produce a unique perception of mechanised time. When a number of mechanised time codes are orchestrated with one another Tinguely highlights that there are conflicting perceptions of time and time codes.

3.1 Problems with Time and Technology in Art in the 1960s

A number of artists in Europe and North America in the 1960s began to experiment with the ways in which time could be explored through art. This was evident in areas both inside and outside of kinetic sculpture. For instance, in his essay titled 'Entropy and the New Monuments in 1969', Robert Smithson reflected on the emerging tendencies that he argued, came to characterise sculpture in the 1960s. These trends indicated shifting perceptions of temporality that became central characteristics of art in the 1960s. He wrote:

> Instead of causing us to remember the past like the old monuments, the new monuments seem to cause us to forget the future. Instead of being made of natural materials, such as marble, granite, plastic, chrome and electric light. They are not built for the ages, but against the ages. They are involved in a systematic reduction of time down to fractions of seconds, rather than in representing the long spaces of centuries. Both past and future are placed into an objective present (p. 11).

According to Smithson, artists such as Donald Judd, Dan Flavin, Robert Morris, and Sol Le Witt were creating works that reassessed traditional time-space relations between the viewer and the work of art. These artists resisted traditional notions of a work of art by creating artworks that moved *with time*, rather than representing time. In this sense, according to Smithson, the use and discussion of duration in sculpture was a means for artists to dematerialise their works and discuss their art from a post-object vantage point.

Smithson echoed an approach to time that Jean Tinguely had already presented a decade earlier. In 1959, at the Institute of Contemporary Arts in London, Tinguely argued against the traditional representation of time in art, and for his audience to become aware of the experience of the present as it unfolds. Tinguely declared:

> The constant of movement, of disintegration, of change and of construction is static. Be constant! Get used to seeing things, ideas and works in their state of ceaseless change. You will live longer. Be permanent by being static! Be part of movement! Only in movement do we find the true essence of things. Today we can no longer believe in permanent laws, defined religions, durable architecture or eternal kingdoms ....We are still very much annoyed by out-of-date notions of time. Please, would you throw away your watches! At least, toss aside the minutes and hours (quoted in Brett 2000, p. 250).

For Rosalind Krauss, Tinguely and Smithson recognised that addressing time in art was an effective way to challenge high-modernism (pp. 201–242) because both artists emphasised the present temporality to harness the attention of audiences in order to engage with art as an experiences rather than representations of time. While Smithson was more concerned with conceptual and minimalist art, Tinguely brought attention to time through his neo-dadaist manifesto and sculptures that moved in front of the viewer.

Smithson and Tinguely's approach coincided with what Time Magazine described as the decade of the 'kinetic kraze' in Europe and North America, which celebrated the motion of mechanical technologies. Such a 'kraze' is evidenced in major international exhibitions such as the Pepsi Pavilion at the Osaka World Fair (1970) presented kinetic artists featured in the exhibition, as forerunners of American art who had represented an aesthetic for technological art to international audiences. Organised by Billy Klüver's *Experiments in Art and Technology* (E.A. T), the pavilion involved a collaboration of artists and engineers who created a range of participatory works, responsive environments, large-scale kinetic installations, virtual reality interfaces and manufactured ecological environments (Klüver 1969, pp. 4–7).

Robert Breer's motorised minimalist fibreglass dome shaped 'floats' slowly glazed around the outskirts of the pavilion. The artist jokingly likened them to emblems of American technological progress: 'I thought, how typically American it would be to actually motorize a Zen garden!' (quoted in Kuo 2010, pp. 214–216). Amongst the high-speed light shows and immersive installations at the pavilion even the most seemingly static works were motorised, with Breer's kinetic sculptures moving at a rate so slow, their movements were barely perceptible as less than two feet per minute. The works exhibited by the artists were publicised not only as an experimentation of technology and art but also at a display for international audiences to understand America as a technologically advanced nation. Similarly to the Kitchen Debate between vice President Richard Nixon and Soviet Premier Nikita Krushchev in 1959, the Pepsi Pavilion was presented to international audiences to reveal an array of technological achievements mastered by Americans, among which kinetic artists like Breer, David Tudor, and Robert Rauschenberg were key representatives (Miller 1998, pp. 20–29).

Considering the presence of kinetic sculpture at the Pepsi Pavilion, the association between kineticism and the modern machine aesthetic was strong in 1970. This has also been indicated by Alvin Toffler, who argued that the 'kinetic image' in art at this time was a celebration of the materiality of industrial machinery. Toffler describes:

> [k]inetic sculpture or constructions crawl, whistle, whine, swing, twitch, rock or pulsate, their light blinking, their magnetic tapes whirling, their plastic, steel, glass and copper components arranging and rearranging themselves into evanescent patterns within a given, though sometimes concealed, framework (p. 157).

The technological utopianism of this time also saw kinetic artists as pioneers of artistic progress. For instance, art historian Jonathan Benthall has reflected that kineticism was 'an attempt to develop a theory from which a new art might grow' (*Science and Technology* p. 101), rather than lamenting early modern art. Leading up to 1970, kinetic sculpture was considered as a popular emerging technological practice that pioneered the experiments between art, science, and technology. Artists working with kineticism were seen as 'space-age artists' who were at the forefront of technology and art (Kuh 1965, pp. 579–609).

The popularity of kinetic art at this time also suggests that it had the potential to foreground anxieties, visualisations, and critiques of social and technological change. For John McHale, the technological expansion after WW2 needed to be expressed and critiqued by artists utilising the same newly domesticated innovations, which provides one explanation of the popularity of kinetic sculpture at this time. As he McHale stated, '[a]ccelerated changes in the human condition require an array of symbolic images of man [sic] which will match up to the requirements of constant change, fleeting impression and a high rate of obsolescence' (in Toffler 1970, p. 156). Therefore, kinesis was not only used to celebrate a machine aesthetic, but also to explore and discuss the sense of accelerated change in modern society.

The 1960s marks a specific nexus in time when the transition between modernity and postmodernity was most evident. This shift brought with it a reconsideration of how time was previously conceived in modernism as a central mechanism for regulating society. For Jameson, this is evidenced by the technological shift that has moved from the speed of the mechanical machine towards the instant nature of the digital age (pp. 8–21). This new transition also prioritised spatial, rather than temporal codes (Schivelbusch; Solnit), thereby marking the 1960s as a decade of temporal turbulence.

If, as Jameson has argued, modernity is in part characterised with an 'obsess[ion] with the secretes of time', postmodernity is as enamored with space ('End of Temporality 697). While some regarded kineticism as the newly emerging dominant form of art that celebrated the movement and interaction with technology in art (Sandberg 1961, p. 161), by the end of the decade kinetic sculpture was considered to be an antiquated modern form of art. Or according to Clement Greenberg, kinetic sculpture was 'ineffective' because mechanical machinery was becoming old without going 'far enough towards the look of non-art, which is presumably an 'inert' look that offer the eye a minimum of 'interesting' (Greenberg 1986, p. 252). The shift from the temporal towards the spatial, as Jameson has identified, explains both the celebration of, and criticism against kineticism, precisely because artists used kinesis to express and explore mechanical rationalisations of temporality.

The computer age that emerged through the aftermath of WW2 is, amongst others, a key contributor to this acceleration that produced what Reinhardt Koselleck has described as a 'peculiar' temporality that differs from the expression of movement earlier in modernity (in Lee 2004, p. xii). For Koselleck, what is different about the acceleration of the everyday within the computer age is the substitution of movement with the *speed* of information systems (Virilio 1984, pp. 101–118). The speed at which actions, information, and feedback can be projected, redistributed, accessed, and processed in the computer age contributes to an increasingly isolated and stationary individual. This is in contrast to the tropes of segregation, repetition, and alienation associated with the spectacle of the modern industrial revolution, that postmodern machines accelerate.

Art historian Pamela Lee has argued that artists in this decade of technological transition created art that was sensitive to the 'bleak prognosis for the condition of time in late modernity' (p. xi), where the spatialisation of time became an indication of a transition from modernity to postmodernity. Consequently, Lee argues, that even though temporality was at one point a site of power, it became 'understood as at once desperate and fatal' (p. xi). This bleak scenario, as Lee has described, is a consequence of newly emerging time-space relationships formed by the domestication of technological expansion and increasing presence of the emerging post-modern society.

Lee argues that time in this period was under threat; A combination of social and political turbulence, fears of international nuclear warfare, cold war tensions, and America's military engagement in Vietnam provided implications that 'the time' as people saw it then, was at risk of great change (pp. 5–34). Combined with the domestication of computing technologies in the public and private spheres in the 1960s, discussions around the changing rationalisations of time through art surfaced. Lee approached a broad range of artists such as Andy Warhol, Pol Bury, Jean Tinguely, and On Kawara and suggested that they were amongst those who explored the phobic relationship to temporality in society. Her art historical emphasis on these artists suggests that not only were the 1960s a time that developed a fear of time described as 'chronophobia', but that this condition has also continued through to contemporary art, theory, and history (pp. 259–308).

Additionally, despite the technological enthusiasm in the 1960s, there was also trepidation towards the consequences of a newly emerging technological society (Ellul 1964; Mumford 1952). Technological militarism was a particularly strong incentive for being skeptical of the expansion, domestication, and appropriation of emerging technologies in society (Jamison and Eyerman 1994), particularly considering the domestication of war atrocities, international nuclear and cold war tensions. The culmination of these concerns and celebrations of technological innovation pinpoints exactly to why Peter Selz described later modernity as a 'turbulent' era, full of disorientation that artists struggled and attempted to engage with (pp. 275–285).

And yet the 1960s was also a unique decade where artists also attempted to expand the modern conceptions of time *with technology*. This attitude is not isolated to kinetic art, but is indicated through the use of real-time in art across a variety of practices (Gaggi 1986). For instance, the popularity of Fluxus art, Allan Kaprow's Happenings, and John Cage's *4'33"* have each been interpreted as widening the role of time in art (Sontag 1966, 24–37; Krauss 1977, pp. 201–242). The popular collaborations between artists and engineers orchestrated by E.A.T, founded by Robert Rauschenberg, Billy Klüver, Fred Waldhauer, and Robert Whitman, also concentrated on the incorporation of interactive computer based systems art, participatory theatre, interactive systems, and the innovative power of art in scientific thought.

The technological expansion of the 1960s also brought new possibilities for the implementation and discussion of artworks that moved in real time with their viewers. Umberto Eco claimed that kinetic sculpture was an appropriate tool for presenting his theory of the 'open work' within the visual arts because they are not works that are fixed in space and time, but unfold with the viewer and their perceptions (Eco 1989, pp. 86–87). For Eco, art that unfolds in real-time enables a specific constellation of perception to be made because the entire work of art cannot be grasped entirely at one time, and demands a perception that builds with duration.

Additionally Willoughby Sharp, the curator of art exhibitions such as *Air Art* (1968), described kinetic art as the prime medium for elaborating on the effect of time when considering art as an 'open work' (Sharp 1968, p. 6). For Sharp the artists in *Air Art*, such as Graham Stevens, Hans Haacke, Pol Bury, Andy Warhol, Les Levine, and Robert Morris form works that 'create time'(6). The durational works in the exhibition expand the perceptions of temporality by unfolding in the same time and space as the viewer, inviting them to be a part of a spectacle in real-time. Sharp continues, '[k]inetic works are more accurately designated as systems [...] Their major function is providing information about how to adapt our extended faculties to technology'(6). For this reason incorporating time and movement into art provided the potential for further creation and new perspectives of time.

The polemics between art, time, and kinesis during the 1960s can be reflected on as a period of uncertainty around the emerging machine aesthetics. While the technological and social changes of this period have warranted descriptions of new conceptions of temporality, peculiar acceleration (Althusser and Balibar 1970), and even phobic conditions to time-based arts (Lee 2004), these arguments form part of a wider debate that expanded the awareness of the role and affect of real-time in art. They have contributed a multiplicity of understandings that renegotiated time-based arts as real-time systems, and open works that responded to the changing perceptions of time in society.

3.2 Against Duration: Art and Objecthood

One of the most widely discussed criticisms against the incorporation of duration in art is Michael Fried's (1998) 'Art and Objecthood'. Published in *Artforum* in 1967, Fried's essay is indicative of an anxiety towards real-time in art, because Fried argued that duration degraded modern conceptions of a work of art. In 'Art and Objecthood', Fried argued that duration in art reduces sculpture to theatricality and prevents it from expressing the complexities of high modernism. Fried's premise is largely formed against Minimalist artists (including of ABC Art, Primary Structures and Specific Objects), and more specifically the works and texts by American artists such as Donald Judd, Robert Morris, and Tony Smith. However, even though his essay is most explicitly directed towards these artists, it also stands as a critique against artists orchestrating a sense of duration in their art across a variety of disciplines.

Fried argues that the works by the above artists are produced with a specific 'objecthood', which is entirely dependent on their reception, interpretation, and presence of the viewer, rather than the artist (pp. 152–153). For Fried this dependency on the viewer to interpret and project meaning onto works reduces the artwork to a mere object, rather than a crafted work of art that has been codified with meaning by the artist. Consequently, Fried argues, there is a theatricality at play between the artist and the viewer where 'the literalist espousal of object hood amounts to nothing other than a plea for a new genre of theatre, and theatre is now the negation of art' (p. 153), to which the works offer very little in return because perception is projected onto the object rather than being drawn from it by the viewer.

Fried continues that theatricality is a great threat to high modernism where 'theatre and theatricality are at war today, not simply with modernist painting (or modernist painting and sculpture), but with art as such', because duration has the potential to collapse the plastic arts into theatrical events (p. 160). Fried is adamant that artists who execute similar presence or performativity, are eroding the quality and status of modern sculpture and therefore must be 'defeated'. He also makes it clear that in order for sculpture to reach what he considers a modern high-art aesthetic complexity as rich as painting, its primary concern should be to abandon any tendency that draws sculpture closer to the temporal, rather than the plastic arts.

Fried argues that a key reason why theatricality in sculpture must be abandoned is because of the way durational works are ontologically unstable and unfold in real-time with the viewer. In these instances, time is not embedded in the form of the work, but is projected as a seemingly endless present that unfolds with the viewer and the work of art. Time of course exists in the plastic arts: time-space relations are codified in the representation of works, and also too, works of art exist in time, as they are subject to decay and destruction. However, time-based arts such as participatory art, as well as minimalist, and kinetic sculpture, are developed with a duration, through which meaning unfolds. Fried is critical of the 'sense of temporality, of time both passing and to come, *simultaneously approaching and receding*, as if apprehended in an infinite perspective', which forms a key concept for his argument against theatricality in sculpture (p. 167). In Fried's perspective, duration '*persists in time* and the presentment of endlessness that, I have been claiming, is central to literality art and theory is essentially a presentment of endless or indefinite *duration*' (p. 167). Fried argues that modern painting and sculpture should be 'wholly manifest', have 'no duration' and be created with a specific representation of time that is embedded in the ontological form of the work.

Such an attack influences the interpretation of kineticism in art. Fried's critique against artists who attempt to redefine and restructure the aesthetic of sculpture by exploring perceptions of temporality also makes an attack on the wider practices of technological art, robotics, media art, participatory art, and happenings at the same time. Rather than having the potential for creating new emerging artistic practices and emphasising new differences between them, Fried falls short and argues that duration in art conflates and reduces the sculpture, to the temporal arts, which is indicative of the confusion and debate that circulated around time-based and technological art.

Susan Sontag has also warned that an uncertainty around to role and affects of time-based art risks either perpetuating an 'intensification of what each art distinctively is', or conflating distinct disciplines and artistic practices into a 'vast behavioural magma or synthesis'; two strains which are irreconcilably opposed to one another (1966, pp. 24–37). Consequently, Sontag argues that time-based plastic art risks being interpreted as a hybrid of practices, rather than an exploration and strengthening of the effects and capabilities of sculpture and theatre in a time of technological flux.

Contrary to Fried and Sontag, Rosalind Krauss has drawn attention to the fact that the exchange between kineticism, sculpture, architecture, and theatre has had a strong tradition in both Europe and the United States in modernity, which only continued in the 1960s (1977, p. 207). To dismiss the history of avant-garde experiments between theatre and sculpture also dismisses the place that kinetic sculpture holds in modern art history. Krauss argues this exact point:

> Fried had asserted that theatricality must work to the detriment of sculpture—muddying the sense of what sculpture uniquely was, depriving it thereby of meaning that was *sculptural*, and depriving it at the time of seriousness. But the sculpture I have just been talking about is predicated on the feeling that what sculpture *was* is insufficient because founded on an idealist myth [sic]. And in trying to find out what sculpture *is*, or what it can be, it has used theatre and its relation to the context of the viewer as a tool to destroy, to investigate, and to reconstruct (Krauss 1977, p. 242).

Krauss points out that the experimentation of interdisciplinary practices, genres, and temporalities arise as methods of reconstruction, rather than weakening practices into hybrid forms like Sontag warned. An important aspect of Krauss' argument lies in her view that kineticism is a tendency that runs parallel to the post-object minimalist art by Donald Judd and Robert Morris, achieving a similar effect in expanding the formal and conceptual relationships between sculpture, time, theatricality, and spectacle (1977, pp. 201–202). In this regard, kineticism was used to move sculpture beyond the materiality of the object, and extend its conceptual and durational boundaries.

In *Expanded Cinema* Gene Youngblood (1975) uses a similar approach to Krauss with respect to the experimentation of time-based arts. Youngblood emphasises the differences between film, theatre, installation, and sculpture that have emerged from the creation of interdisciplinary art. Like Krauss, Youngblood demonstrates that it is only through the dissolution of traditional distinctions of art that new artistic genres emerge and investigate the limits of each discipline. For instance, the incorporation of duration raises questions around intermedia theatre, projected environments, and expanded cinema as distinctive aesthetic forms. To quote Youngblood:

> Although intermedia theatre draws individually from theatre and cinema, in the final analysis it is neither. Whatever divisions may exist between the two media are not necessarily 'bridged', but rather are orchestrated as harmonic opposites in an overall

> synaesthetic experience. Intermedia theatre is not a 'play' or a 'movie'; and although it contains elements of both, even those elements are not representative of the respective traditional genres (p. 365).

Therefore new disciplines emerged through the hybrid experiments of existing disciplines. Sontag, Youngblood, and Krauss understand that time-based arts can collapse the distinctions between aesthetic practices and also reconstruct traditional modes of representation. Therefore, it is through the incorporation of duration, an exploration of the differences between sculpture, theatre, cinema and performance emerged.

As noted earlier in this chapter, the debate that has surrounded 'Art and Objecthood' since its publication indicates that Fried's disregard for the use and expression of duration is symptomatic of the temporal turbulence in the 1960s. Fried's argument in 'Art and Objecthood' is problematic for artists experimenting with kineticism as an expression of time because, so Fried argues, duration reduces sculpture to a spectacle of theatricality, which resists a modern definition of a work of art. The shift towards expressing time with movement is another indication of the technological change and new means for conceiving and experiencing time in the everyday, in a manner that speaks to the peculiar relationship to time in a period of technological change.

3.3 *Directions in Kinetic Art* at the Berkeley Museum (1966)

For those artists who were experimenting with actual movement in their practices in the 1960s represents a desire to understand the changing conceptions of time at the time. While some exhibitions of kinetic sculpture focused on kineticism as an exploration of a modern industrial machine aesthetic such as *The Machine as Seen at the End of the Age of Mechanical Reproduction*, others such as *Le Mouvement* at the Denise Rene Gallery (1955/1967), *Slow Motion,* Rutgers University (1967), *Kineticism: Systems Sculpture in Environmental Situations* at the Museum of Arts and Science in Mexico City (1968), *Place and Process* (1968), *Kinetic Environments I* and *II* (1967), and *Air Art,* New York (1968) explicitly presented artists who were interested in the use of movement to influence perceptions of time. For instance Yaacov Agam, an artist who had already participated in exhibitions such as *Le Mouvement*, used mechanical movement to explore time as a constant transformation to highlight that '[w]e are different from what we were three months ago, and in three minutes more, we will again be different. […] The image appears and disappears, but nothing is retained' (in Toffler 1970, pp. 155–158). Rather than using mechanical media to articulate movement, speed and acceleration, Agam is focused on modern industrial media as a tool for exploring the perception of time as a constant duration.

Similar to Agam's approach to time, the exhibition *Directions in Kinetic Sculpture* at the Berkeley Museum of Art (1966) exhibited a generation of kinetic artists who were interested in the role and effects of movement as an expression of the present. One year prior to Fried's 'Art and Objecthood' the exhibition was certainly not the first to feature a generation of young artists working with mechanical movement,[1] however it was one of the first to openly discuss kineticism in a self-reflexive manner. The public reception of the exhibition was extremely positive with over 80,000 attendees, which indicated that kinetic sculpture had a future as an accessible modern art form.

In the exhibition catalogue, documentary material, and transcripts from the exhibition and symposium, the artists and curator Peter Selz discussed the roles and effects of movement in sculpture. Rather than showing an historical survey of kinetic sculpture, Selz curated the exhibition to feature 15 artists who he saw was representing the post-war generation of artists experimenting with new technological media in art (Selz 1985, pp. 287–302). The artists featured were predominantly from Europe and North America and included, amongst others, Fletchor Benton, Robert Breer, Pol Bury, Jean Tinguely, and Len Lye.[2] The exhibition presented kinetic sculpture and installation through four dominant themes: theatrical spectacle, carrier of natural energy (particularly wind and water), of making the invisible visible, and discovering the formal capacity of mechanical movement and media. Within these themes, kineticism was presented as a practice that intersected with the physical sciences, neo-dada aesthetics, conceptual and technological art (Selz 1985, p. 288).

The shifting conceptions and perceptions of temporality were persistent subjects throughout the symposium, particularly because many of the artists expressed an interest in using kinesis to explore durational rhythms (Selz 1985, pp. 288–302). For instance, in the exhibition catalogue, Selz stressed the relationship between the kinetic dynamism and Henri Bergson's duration as a continual process of change in order to focus on the entropic, unpredictable and productive nature of the present as it unfolds. Selz emphasised Bergson's concept of duration in *Time and Free Will*, where the movements of pure duration are 'internal and heterogenous to one another, and in pure duration a cause cannot repeat its effect since it will never repeat itself' (Selz 1985, p. 2).

---

[1]Other exhibitions include: *Le Mouvement* (1955/1967), *Nul* (1965) Stedeljik Museum, Amsterdam, *Kinetische Kunst* (1960), Kunstgewerbmeuseum, Zürich; *Bewogen Beweging*, (1961) Stedelijk Museumm Amsterdam; *Movement in Art* (1961), Howard Wise Gallery, Cleveland; *On the Move* (1964), Howard Wise Gallery, New York; *Movement* (1965), Hanover Gallery, London; *Kinetic and Optic Art Today* (1965), Albright-Knox Art Gallery, Buffalo; *Art and Movement* (1965), Royal Scottish Academy, Edinburgh; Glasgow Art Gallery, Glasgow; *Art et Mouvement* (1965), Musée de Tel Aviv, Tel Aviv; *Kinetic Art*, (1965), Galerie 20, Amsterdam; *Two Kinetic Sculptors: Schöffer and Tinguely* (1965), Jewish Museum, New York.

[2]The full list of artists: Fletchor Benton, Davide Boriani, Robert Breer, Pol Bury, Gianni Colombo, Gerhard von Graevenitz, Hans Haacke, Harry Kramer, Len Lye, Heinz Mack, Charles Mattox, George Rickey, Takis, Jean Tinguely, Yvaral.

This description of duration, quoted from Bergson's *Time and Free Will*, suggests that there are two kinds of multiplicities in time that relate to one another. Firstly, there is an ontological quantity (space), and secondly a multiplicity of states of consciousness as a process of qualitative transformation that cannot be quantified (pure duration). For Bergson, this pure duration is the time that we experience subjectively. Like our awareness of time when we experience it, time also moves in rhythms—waning and intensifying as it passes—an experience that can never be truly represented by media (Bergson 1913, pp. 321–323).

Selz claimed that Bergson's understanding was crucial to the exhibition because artists were exploring a variety of technological, natural, and communicative movement to expand the ways in which we think of time (Rickey 'Introduction', *Directions in Kinetic Sculpture* pp. 1–11). New technological media in motion therefore had the potential to create new sensations and perceptions of duration. Selz's reference to Bergson is both a dedication and complication of Bergson's approach to time and technological representation because mechanical movement was used to present, rather than represent or abstract, the sensation of pure duration in the exhibition. It is important to note that for Bergson duration can never be exactly reproduced or represented by mechanical media (Bergson 1913, pp. 322–323).

The different approaches to movement and time were also openly discussed between exhibiting artists such as Len Lye and George Rickey:

> *Lye*: George, do you think of time when you're composing a figure of motion? Do you consider, well, now I've got a thing I want to be about a minute long?
>
> *Rickey*: Well, I certainly consider the velocity, which is an expression of time.
>
> *Lye*: Yeah, but not in terms of 'tick-tick'.
>
> *Rickey*: A little slower than that. But I certainly think of it—I think of it all the time…I think rather of time as being a material that can be ordered in an exactly equivalent way that space is a material that can be ordered (Selz 1985, pp. 293–294).

Here Rickey is more concerned with the orchestration of speed and velocity, while Lye, like Bergson, brings attention to the mechanical rationalisations of time as a constraint for creating works that perform movement and express duration. Similarly, Jean Tinguely used movement to create, rather than capture time, and present a sense of endlessness through his idiosyncratic non-machines. This is made explicit when he explained, 'I began using movement simply as a method of re-creation. It was a way of changing the image to make it infinite' (in Le Jeune).

Discussing the affects of kinetic art enabled the symposium to explore the ontological implications of kineticism and attempted to separate it from other disciplines that incorporate movement. The symposium and exhibition therefore emphasised kineticism as a catalyst for expanding the role of time in art, and creating new compositions and experiences of time. Selz addressed the panel as follows:

> The question that arises is: is this kinetic sculpture a new movement? Is it a specific movement , such as, let's say, movements which are perfectly clear in one's mind, like Cubism or Surrealism, which had their beginning, their climax, their decline? […] Are we

dealing here with a specific kind of movement in art, or is it an art form, which has nothing to do with a specific movement? (Selz 1985, p. 297).

This question may have been designed to encourage debate and discussion between artists at the symposium. However it also indicates that there was a distinct attempt to insert kinetic sculpture into the schema of modern art, rather than discussing how kinesis might be used to help redefine the parameters of modernism. Therefore, unlike Fried's resistance to time-based art in 'Art and Objecthood', *Directions in Kinetic Art* used movement and temporality as mechanisms to strengthen and explore kineticism as a discipline, rather than as elements that reduce sculpture to theatricality.

3.4 Orchestrating Time: Jean Tinguely's Auto-Destructive Machines

Selz's decision to include Jean Tinguely in *Directions in Kinetic Sculpture* was informed by the unique calibre, breadth, and humour that Tinguely had already been commended for using in his kinetic sculpture. Tinguely is now most recognised for his neo-realist, neo-dada antagonism and auto self-destructive performances in both Europe and the United States (Carrick 2010, pp. 67–102), and has since been referred to as the 'father of kinetic art' (Pakesch p. 64). By 1965, Tinguely had already collaborated with Yves Klein, Nikki de Saint-Phalle, John Cage, and Robert Rauschenberg and the *Nouveaux Réalistes* (New Realists) and had already exhibited his kinetic works with a variety of artists from Europe and the United States. Such exhibitions include *Le Mouvement* (1955), at the Denise Rene Gallery in Paris, and *Bewogen Beweging* (translated as *Art in Movement*), curated by Pontus Hultén at the Stedeljik Museum in Amsterdam in 1961.

In a similar vein to the film *Modern Times,* which starred Charlie Chaplin, Tinguely presents a rationale for modern machinery dominating and regulating the times of labour and leisure. However, to only consider the artist's oeuvre as a critique of the modern machine aesthetic suggests that the contemporary exhibition of Tinguely's work fetishises modern antiquity and neo-dadaist sensibilities common to the *Nouveaux Réalistes.* Another pervasive element of many of Tinguely's works directly with the present temporality as a mode for inciting action, harnessing attention, and manipulating perceptions of duration passing through it.

The four works of Tinguely's that featured in *Directions of Kinetic Sculpture* were, *Radio Sculpture with Feather* (1962), *Radio Sculpture Number 5* (1962), *Suzuki* (1963) and *M.K. III* (1964). These encapsulate Tinguely's critique on industrial technology and are both assemblages of recycled machine parts and transmit live radio sounds out into the gallery space. Rather than as an uninterrupted streaming the radio transmission is fragmented and fades in and out of range delivering 'dismembered' soundscapes that are juxtaposed against the repetitive motions of the mechanical components of the work. *Radio Sculpture with Feather* is a standing metal box comprised of mechanical parts from different machines and is constructed with an aerial protruding from its head and a feather attached at its end. At repetitive intervals, the sculpture kicks into motion to make the aerial wag wildly in the air above it. Like *Radio Sculpture with Feather, Radio Sculpture Number 5* also moves in constant repetitive steps that propel the machine into motion without causing an effect: a mechanical performance that is void of producing something other than its movements.

Suzuki (1963) and *M.K. III* (1964) are also key examples of Tinguely's machines fitted with a patchwork of belts and motor parts. Unlike Tinguely's *méta matics* and self-destructive works, their movement and production is limited to bursts of repetitive actions and rest. *Suzuki* is assembled in a contorted arrangement that balances in a top-heavy pose on its mount, while *M.K. III* is composed with belts that run in continuous loops like empty conveyor belts. By comparison to his auto-destructive performances and antagonist works, these pieces move in a calmer and more regular fashion with their belts looping around and through the sculpture.

These four artworks do little to display the breadth of the artist's work. If the exhibition aimed to showcase the emerging artists at the time, and the directions in which they were leading kinetic art, Selz curated a conservative image of Tinguely's exploration of movement with technology in art. *Directions of Kinetic Art* did not proclaim to produce a comprehensive range of kinetic artworks at the time, but rather paid homage to the artists who were attempting to build, produce, and explore the relationship between kinesis and late modernity. In order to gauge the breadth of the artist's use of movement to orchestrate a specific experience of temporality, it is therefore important to expand this analysis out to other works by Tinguely.

In addition to a critique of modern labour and consumption, Tinguely also used spectacles of auto-destructive performances to concentrate on the uneasy relationships his audiences form with the present temporality. His mechanical non-machine machines such as *Homage to New York* and *A Study for an End of the World I* and *II*, unfolded in time in an entropic and unpredictable to contrast with the rationalised and repetitive movements of modern industrial machinery in the workforce.

From an early age Tinguely had a fraught relationship with how time for work and leisure were regulated in society, and before he became an artist he was fired from a department store for ripping the store clock off the wall before storming out in an act of defiance against the social expectation that time equals money. Later as an artist this frustration came out in his artwork. For instance, for the *Concert of Seven Pictures* in Düsseldorf (1959) Tinguely distributed 15,000 copies of his manifesto titled, *Für Statik* (For Static) throughout the town via an aircraft. The manifesto called for the honour of actual movement and encouraged the production of gestures, motion, and time, in real time, and to abandon forms of static representation in art. The manifesto proclaims:

> Forget hours, seconds, and minutes. Accept instability. Live in Time. Be static—with movement . For a static of the present movement . […] Live in the present, live once more

in time and by Time—for a wonderful and absolute reality.... Stop painting 'time'...live in time and according to time for a wonderful and absolute reality (J Tinguely, *Für Statik 1959*, in Danchev 2011, pp. 336–337).

Tinguely's manifesto highlights the central principles behind the artist's interest in time in relation to art, and echoes early European avant-garde manifestos such as Naum Gabo's *Realistic Manifesto* by calling an attention to and privileging of the present above other temporal modes. Tinguely, however, adopts a more critical tone against society for not living in the present, and instead being distracted by mechanised rationalisations of time.

In *Für Statik* the artist asks the viewers art to confront the regulation of time as a mechanically induced regime. This is because for Tinguely, to be aware of the passing present is to also resist mechanical rationalisations of time. Tinguely's interest with time, technology, and movement was accompanied by many others in the arts. For instance Willem Sandberg, the director of the Stedeljik Museum in Amsterdam stated after *Bewogen Beweging* that kinetic art was then 'the art of our time' (in Lee 2004, p. 103) because it played out the tension between art and technology that were a key social concern at the time.

Prior to *Directions in Kinetic Art*, Peter Selz was the curator for painting and sculpture at the Museum of Modern Art (MoMA) in New York. In February 1960 Selz commissioned Tinguely to create his auto-destructive machine *Homage to New York*, to be performed in the MoMA sculpture garden. Compiled from found scrap metal and junk machinery, pieces of the work were collected from rubbish dumps and mechanical dealers in Newark, Summit, and New York and included found objects including eight bicycle wheels, a bassinet, washing machine parts, a weather balloon, piano, radio, American flag and a child's toilet. *Homage to New York* was also constructed with multiple sections, each painted in white and programmed to start in motion at its own allocated time, which was intended to enable the entire sculpture to slowly turn in separate sections and build into a crescendo of mechanical motion, and lead up to its self-destruction.

On March 17 1960, the only performance of *Homage to New York* was set into motion, spurring a cacophony of sounds, smells, and mechanical movements . At one stage a piano played while glass bottles were cast from the top, smashed on the ground and spread nauseating smells while a child's go-cart paced back and forth in front of the work. The performance was one of industrial mechanical excess as indicated by Tinguely who commented that: '[t]he machines which we build today produce much more than we can possible consume. I solve this problem of abundance in my own way!' (in Carrick 2010, p. 72). Tinguely emphasised this by making a work that did not produce anything other than motion; a re-staged spectacle of abundance, from abundance.

It is therefore no surprise that *Homage to New York* is remembered as a critique of the physical abundance of consumer materials in capitalist society (Stiles and Selz 1996, p. 402). However, Tinguely's kinetic performance was also an orchestration of duration, anticipation, and suspense to confront his audience with a new awareness of time. For the experience of the performance itself one can only draw from the documents, photographs, and film recorded by Robert Breer, as well as first hand accounts from the audience at MoMA that night that witnessed the sole performance. Billy Klüver, an engineer from Bell Laboratories who assisted Tinguely in the compilation and construction of *Homage to New York* reflected on the performance as an experience riddled with interruption, unplanned accidents, and suspense for both the unknowable and yet programmed mechanical self-destruction. In various sections of Klüver's account he emphasises the anxiety that was felt while waiting for the work to unfold, and then destroy itself:

> Not once did we go over everything and check it…the arm he had worked on with perfection did not work…Something was wrong with it, it was winding too slowly…In the eighteenth minute, the fire extinguisher in the piano was supposed to go off. It didn't.…The whole machine was somewhat sick after the bad handling in transport, and it fell over after only a few minutes.…After three minutes, the longest in my life, they finally began to put out the fire.…At this point Jean and I were almost desperate (pp. 74–77).

Klüver's reaction to the unpredictable performance is described as a moment of excruciating suspense. The time that the sculpture took to inevitably self-destruct painfully stretched out and tested the threshold between expectation and actualisation. In this moment, time waned and slowed because the sculpture moved in ways that most everyday machines do not, and destructed in an uncertain and chaotic manner. Klüver's reaction to the movement of the work emphasises Tinguely's intention to manipulate a sense of time—one that draws out, prolongs, and becomes excruciating as a result of anticipation.

Klüver's experience of waiting was an intensive, rather than extensive sensation. Throughout the performance Klüver was not only sensitive to his perception of time, but became arrested in suspense as he waited. This is because for Klüver to wait during *Homage to New York*, was to be seized by it, and drew attention to its uncontrollable and unpredictable nature. To wait during the final three minutes before the fire was finally extinguished, Klüver's perception of the passing of time slowed to what felt like, an endless state. In a similar respect to the reactions to Tinguely's works at *Directions in Kinetic Art*, Klüver and fellow audience members were void of agency other than being made aware of their impatience, uneasiness, and uselessness. Through Klüver, Tinguely's work created an unease with the present temporality because of the unpredictability and uncertainty which felt endless. The uneasiness that Tinguely created not only existed because the performance heightened an awareness of duration by audience members but also because Tinguely brought forward and made visible movement in time, in art.

This perspective is also affirmed by Selz who stated that a key effect of Tinguely's kineticism is his ability to create a specific relationship between mechanical movement and time that is specific to kinetic sculptural media: "Jean Tinguely's experiments are works of art in which time, movement, and gesture are demonstrated—not merely evoked" (in Lee 2004, p. 137), which is consistent with Tinguely's earlier claims in *Für Statik.* We might say therefore the Tinguely used movement in art to present and *persuade* sensations of time, rather than represent them. Such persuasion is achieved in *Homage to New York* by bringing attention to the tensions between rationalised time and time as a state of transformation. Some of the sections within the work delayed, while some began earlier than they were supposed to. For instance, a radio piece played, but the audience could not hear it above the noise of the machine, while some objects rolled in the wrong direction and made paintings out of beer rather than paint, and the overall performance went on for three times longer than Tinguely had anticipated.

Homage to New York also acted as an allegorical performance that celebrated the gradual emerging antiquity of modern rationalisations of time. While some have understood the exploding self-destructive machine as a commemoration for the end of modern industrial machinery (Reichardt 1987 369; Hoptman 2000, pp. 8–9), Tinguely also demonstrates that in the wake of technological transition, the conceptions of time-space relationships are also changing within this greater technological and social shift. To create a mechanical landscape out of an abundance of recycled and rejected machinery is not only a warning for the increasing antiquity of modernity, but also acts as a signpost for the new ways time is spent, divided, expressed, and perceived.

This temporal experience is primarily characterised by waiting for change that does not happen when you expect it (and also happens when you do not expect it). Although Klüver lived through the performance, his perception of time was paced by his own hesitancy and passive inaction. Henri Bergson too, has described that the hesitancy that is felt during the present, in the process of actualising from the virtual, as a felt hesitancy that is associated with the unpredictable and productive nature of the present. Therefore we might say that Klüver's hesitancy is a reaction that is indicative of the anticipation and anxiety about change that occurs later than expected, and outside of any control.

Tinguely's performance at MoMA directly resonates with Fernard Léger's description of modern spectacles. Léger wrote in his essay titled 'The Spectacle', which was published soon after his film *Ballet Mécanique* in 1924, that '[t]here is in the origin of the modern spectacle…the shock of the surprise effect'(Crary 2002, p. 463). Tinguely uses movement to create a similar kind of spectacle of movement that arrests its audience with shock and attention through kinesis in his art. Although the work inevitably destructed, Klüver's shock originated with the way the machine moved unpredictably. Tinguely's cacophony of simultaneous motions in *Homage to New York* also contributes to an open and distributed experience of time. Rather than coming together to form a total work of art, Tinguely uses mechanical movement to distribute the work across time.

Through Tinguely we can understand that perceptions can be heightened when attention to the temporal present is harnessed in the moment. Rather than considering that perception is an act of the subject drawing perceptions away from the object, perception is distributed and accumulated over time. The perception of the spectacle within the present has an unavoidable intensity, as Tinguely suggests: 'only in movement do we find the true essence of things' (p. 119). Tinguely emphasises that modern industrial mechanical movement has the productive potential to pull in an awareness of duration, rather than being a technological tool for abstraction and reproduction. While *Homage to New York* may have achieved its desired effect, as evidenced by Klüver's reaction, Tinguely's subsequent works highlighted the incompatible time codes that exist between technologies, rather than tossing aside minutes and hours, and bringing awareness back to the passing of real time.

Following *Homage to New York* in 1960 Tinguely created two more self-destructive works titled, *A Study for an End of the World* (numbered I (1961) and II (1962)). The first was installed at the Louisiana Museum of Modern art in Humlebaeck, north of Copenhagen in Denmark, at the exhibition entitled *Movement in Art*. The second was commissioned by NBC's television series *David Brinkley's Journal,* and was performed in the Nevada desert of Las Vegas in 1962. While *Study for an End of the World I* performed similarly to *Homage to New York*, the second *Study for an End of the World* is significant for proposing a critique against modern industrial spectacles of time and production, as well as the domestication of television and telecommunications that became emblems of the middle class modern American lifestyle.

Like *Homage to New York* the parts that formed *A Study for an End of the World II* were sourced from nearby rubbish tips in Las Vegas and controlled by Tinguely from afar. It also comprised several sections that were programmed to set into motion at a specific time before entirely exploding. Even though *A Study for an End of the World II* was created with more sophisticated control mechanisms than *Homage to New York*, the performance was disrupted by more technical glitches.

Much to the artist's delight, the timed sections delayed their performances, and the entire destruction took an hour longer than it had been programmed to. He responded to these glitches by commenting humorously that, '[i]t's not to be expected that the end of the world will be exactly as it's been imagined' (Hultén p. 119). The spectacle of *A Study for an End of the World* differs greatly to *Homage to New York*, precisely because of the conflicting time schedules that the later performance was subject to. Even though *Homage to New York* was recorded with media photographers and captured on film by animator and kinetic sculptor Robert Breer, all documentary media was used to capture the movement and time rhythms of the sculpture itself. In contrast, *A Study for an End of the World* highlighted the incompatible time codes between modern industrial machinery and the newly domesticated telecommunications technology that ran at different schedules.

Tinguely highlighted the pressure and failure of modern industrial machinery and rudimentary robotics to keep up with telecommunication in *A Study for an End of the World II*. The televised documentation of the performance prevented the viewers from witnessing the actual pace and rhythm of the work's destruction. The spectacle was edited, spliced, and re-presented according to NBC's own image of time that prevented a confrontation with the unfolding present. While the telecommunications technologies in this performance distorted the experience and duration of the performance, *A Study of the End of the World II* highlights the intersections and incompatibilities between various technological regulations of time, and therefore develops an appropriation of the modern spectacle to *Homage to New York*.

As Charlie Gere has argued, modern art has throughout its history, continually formed critiques and reactions to the increasing speed and acceleration of technology in the modern era (Gere 2006, p. 13). The changing conceptions of temporality are central to the critique and representation of modern acceleration. Lewis Mumford argued this precisely when he stated that rather than the steam engine or any other modern invention the clock was the most pervasive invention of modernity (p. 8), because the regulation of time mobilised and synchronised the private and public spheres more so than any other modern industrial technology. The regulation of time that commanded the division of labour and leisure implemented a regime of time that influenced the greatest social change of modernity. As Jacques Ellul similarly identified, all life including the facets of work and leisure, and indeed the perception of change and motion continues to be approached and governed by the clock (pp. 329–330). With Mumford and Elull in mind, Tinguely's performances can be used to highlight that temporality undergoes a conceptual shift within the emergence of each technological age. In his case, the time codes between industrial machinery, which are fixed in time and space, and the edited rhythms of telecommunications, dislodge and reset the rhythms of the work's destruction.

The spectacle that Tinguely orchestrated in *A Study for an End of the World II* differs greatly from *Homage to New York*, precisely because of the conflicting time schedules that the later performance was subject to. In effect, Tinguely highlights how different kinds of technology ascribe to their own rationalised time code. Even though *Homage to New York* was recorded by media photographers and captured on film by animator/kinetic sculptor, Robert Breer, all documentary media was used with the aim of capturing the movement and time rhythms of the sculpture itself. *A Study for an End of the World II*, highlights the incompatible schedules between modern industrial machinery and the newly domesticated telecommunications technology. This is emphasised by the pressure and failure of the performance to keep in time with the pace of the time codes that pace telecommunication. *A Study for an End of the World II* was marked with an expectation around the pace and synchronicity of various technologies coming together, that was not prevalent in *Homage to New York*. Broadcast out to viewers, the televised documentation of the performance prevented its audience from witnessing the actual pace and rhythm of the work's destruction.

3.5 Conclusion: Technological Expansion and New Durations

This chapter addressed the temporal turbulence of the 1960s through Michael Fried's essay 'Art and Objecthood', the exhibition *Directions in Kinetic Sculpture,* and the auto-destructive kinetic works by Jean Tinguely. These three examples each address temporality as a polemic in art during this decade due to the changes in domesticated technologies in western societies. While Fried argued that the incorporation of duration in the plastic arts reduces sculpture to theatricality, Selz and the artists in *Directions in Kinetic Sculpture,* and Tinguely's works demonstrate that a key defining affect of kineticism in art is the ability to express and explore a variety conceptions of temporality and time codes. Additionally, I have argued that Tinguely produced works that confronted his audiences with temporal systems that highlight the tensions and conflicts that emerge between the movement of modern machinery and the transmission of telecommunication technologies.

It is through *Homage to New York* specifically that, when attention is brought towards the present temporality, the uneasy and unpredictable nature of the present is emphasised. Tinguely articulates the transition from an industrial age towards a new mechanical age by confronting his audiences with their awareness of duration, as well as the manipulation of time by mechanical and communication technologies. His kinetic performances therefore build new perceptions of time, and provide openings for how temporality can be discussed in new ways.

When considering the durational effects of Tinguely's auto-destructive performances in relation to the discussions and anxieties about temporality in art in the 1960s, it is evident that Tinguely and the artists in *Directions in Kinetic Sculpture* were interested in expanding the sensations and perceptions of time in relation to technology and sculpture, as well as highlighting the mechanised rationalisations of time demanded by the industrial age. Fried's essay 'Art and Objecthood' in 1967, is still remembered as a key document in contemporary art history that argued against the experimentation with time in art, in an attempt to preserve the ideals of high-modernism. However, I have highlighted that Fried fails to acknowledge the social and technological contexts in which time-based art were made: one that was undergoing uncertainty during technological change in society, through which new forms or art could be made.

The polemics addressed in this chapter also give rise to new questions concerning the contemporary art history of avant-garde kineticism. Considering that the kinetic art in the 1960s is marked by both popularity and criticism as a popular form of technological art, the following chapter addresses Jack Burnham's 'Systems Esthetics' as a key contributor to the assumption that kineticism refers solely to the movement of a machine aesthetic. The domestication of computing technologies, as well as the popularity of Ludwig von Bertalanffy's *General Systems Theory*, was used to emphasise movement as a systematic process, rather than a movement of mechanical or non-mechanical form. The popularity of Burnham's appropriation of general systems theory in contemporary media art history and theory is problematic because, as I will argue, his approach in 'Systems Esthetics' is dependent on a disregard for kinetic movement as a means for expressing and exploring systems theory.

References

Althusser L, Balibar É (1970) Reading capital (trans. B. Brewster). NLB, New York
Bergson H (1913) Creative evolution (trans. Mitchell A). Macmillan, London
Bois YA (2000) Force fields: phases of the kinetic. Artforum 39(3)
Brett G (2000) Force fields phases of the kinetic. Hayward Gallery, London
Carrick J (2010) Nouveau Réalisme, 1960s France, and the Neo Avant-Garde: Topographies of Chance and Return. Ashgate, Farnham, Surrey
Crary J (2002) Spectacle, attention, counter-memory. In: McDonough T (ed) Guy debord and the situationist international. The MIT Press, Cambridge, Massachusetts and London
Danchev A (ed) (2011) 100 Artists' Manifestos; From the futurists to the stuckists. Penguin, London
Eco U (1989) The open work. Harvard University Press, Cambridge Mass
Ellul J (1964) The technological society (trans. Wilkinson J). Random House, New York
Fried M (1998) 'Art And Objecthood', Artforum, 1967; reprinted in Art and Objecthood. University of Chicago Press, Chicago
Gaggi S (1986) Sculpture, theatre and art performance: notes on the convergence of the arts. Leonardo 19(1):45–52
Gere C (2006) Art technology time. Berg, Oxford
Greenberg C (1986) Modernism with a vengeance. University of Chicago, Chicago
Hoptman L (2000) What is useless science? MoMA 3(5):6–9
Jameson F (1984) Periodizing the 60s. Social Text 9(10):178–209. doi: 1.http://www.jstor.org/stable/466541
Jamison A, Eyerman R (1994) Seeds of the sixties. University of California Press, Berkeley
Klüver B (1969) Experiments in art and technology. Members Newslett 3:4–7
Kuh K (1965) Recent kinetic art. In: Kepes G (ed) Nature and art of motion. Studio Vista, New York
Kuo M (2010) Everything goes. Artforum 49(3):214–216
Krauss R (1977) Passages in modern sculpture. Thames and Hudson, London
Lee P (2004) Chronophobia: On time in the art of the 1960s. MIT Press, Cambridge
Miller P (1998) The engineer as catalyst: billy Klüver on working with artists. IEEE Spectr 35 (7):20–29
Mumford L (1952) Art and technics. Columbia University Press, New York
Palmer D (2008) Participatory media: visual culture in real time. VDM Verlag, Saarbrücken, Germany
Reichardt J (1987) Machines and Art', Leonardo 20th Anniversary special issue: Art of the futur. Future Art 20(4):367–372
Sandberg W (1961) For movement's sake. Newsweek, March 13
Sharp W (1968) Air art. Kineticism Press, New York
Selz P (1985) Art in a turbulent Era. UMI Research Press, Ann Arbor, Michigan
Smithson R (1996) Entropy and the new monuments. Robert Smithson: the collected writings. University of California Press, Berkeley, pp 10–23
Sontag S (1966) Film and theatre. Tulane Drama Rev 11(1):24–37
Stiles K, Selz P (eds) (1996) Theories and documents of contemporary art: a sourcebook of artists' writings. University of California, Berkeley
Tinguely J (1973) Dynamo Tinguely' In: Piene O, Mack H (eds.) 1970 Zero (trans. Beckman H). Cambridge, Massachusetts, MIT Press. [originally published in Germany in 1958 and 1961]
Toffler A (1970) Future shock. Bodley Head, London
Virilio P (1984) The lost dimension. D. Moshenberg, Semiotext(e), New York
Youngblood G (1975) Expanded cinema. Dutton, New York

Chapter 4
Systems Aesthetics: A Key Polemic in Contemporary Kinetic Art History

> *This is a shift from being to becoming. Kinetic works reflect this shift since kinetic works refute static space. They destroy lineal time. Kinetic works do not occupy space, they create space. Kinetic works do not contain time, they create time. Kinetic works so not interpret reality, they are reality.*
>
> (Sharp 1968, p. 4)

Abstract This chapter positions Jack Burnham's writing on modern sculpture as critical to common misconceptions around kinetic sculpture and installation. Burnham argued that artists working with mechanical movement failed to remain relevant to the technological postmodern aesthetics that emerged in the 1960s. Burnham's perspective is a key contributor to the assumption that kinetic art is an antiquated precursor to other investigations in art and technology. For instance in 'Systems Esthetics', 'Real Time Systems' and *Beyond Modern Sculpture* Burnham argued that the popularity of movement in real time with the viewer became superseded by conceptual investigations such as the use of systems theory in art. However, I argue that artists such as Hans Haacke depended on the use of movement in order to investigate systems theory. Therefore, while Burnham's antipathy towards kineticism separated it from the emerging systems aesthetics in art, Haacke explicitly emphasised the importance of movement and form to connect media with specific conceptual messages

In *Revolutions—Forms that Turn* Carolyn Christov-Bakargiev helped to reify the notion that the modern machine aesthetic is also a contemporary issue. Similarly, *Force Fields: Phases of the Kinetic* was publicised as Guy Brett's attempt to 'excavate' a forgotten experimentation with mechanical movement and present it to contemporary audiences to suggest that approaches to movement and time in art have changed since the late 1960s (Borja-Villel in Brett 2000, p. 7). While both Brett and Christov-Bakargiev turned to early kinetic sculpture to reflect on contemporaneity in contemporary society, they did so because art history, theory, and criticism have previously concentrated on separating mechanical media and modern art from postmodern and contemporary aesthetics.

The way kinetic sculpture in art history has been framed is largely influenced by Jack Burnham's seminal essay entitled 'Systems Esthetics' (1968b). Here, Burnham

© Springer Nature Singapore Pte Ltd. 2017
C. Chau, *Movement, Time, Technology, and Art*, Springer Series on Cultural Computing, DOI 10.1007/978-981-10-4705-3_4

argues that artists creating kinetic sculpture in the 1960s failed to be relevant to the emerging postmodern aesthetics, and thus were left behind. Consequently the popularity of Burnham's critique contributed to the assumption that kineticism is a purely modern concern, which has since only acted as a precursor to media art practices that have emerged since the 1970s.

During the final years of America's 'kinetic kraze' ('Styles' 1966, pp. 66–69) Jack Burnham published two seminal essays in *Artforum*, 'Systems Esthetics' (1968b) and 'Real Time Systems' (1969), and also his book *Beyond Modern Sculpture: The Effects of Science and Technology on the Sculpture of this Century* (1968a). Across these three publications, Burnham considered sculpture in the 1960s as 'a unique child in a unique age' (1968a, p. 4) due to the popularity of modern 'pseudo machines' like kinetic sculpture, which, he considered failed to reconstruct or mimetically perform life successfully (1968a, p. 5). Across these texts Burnham also argued for the understandings of Ludwig von Bertalanffy's *General Systems Theory* as a potential model for approaching the intersections between art and science in the emerging technological age. Consequently Burnham regarded kineticism as an unfulfilled practice, in which its artists had failed to assert their art within the frameworks of emerging systems theory and postmodern aesthetics.

This chapter argues that the popularity of Burnham's 'Systems Esthetics' has been influential in framing kineticism as an antiquated modernist machinic practice that predominantly serves as a precursor to contemporary media arts. That is, even though many pieces of contemporary art move in real time and space with the viewer, the term 'kineticism' has come to be associated with modern antiquity. While Burnham uses systems theory to exclude the effect of movement from critical discourse, I argue in this chapter that artists such as Hans Haacke—whose work Burnham depended on to demonstrate the operation of systems aesthetics in art—also emphasised the form, function, and movement of his early sculptural systems. Contrary to Burnham's perspective, which defines kinetic sculpture solely according to the movement of mechanical form, Haacke approaches kinesis as an orchestration of movement that is used to heighten sensitivity of viewers and make them acutely aware of the present unfolding in time, where there is a perceptual edge between what is and with is becoming.

Burnham's argument is a key influence on the contemporary art historical associations of kinetic sculpture, which is a polemic for artists and institutions producing and exhibiting kinetic works, because they are quickly regarded as referring to a modern aesthetic. As I discussed in the introduction of this book, exhibitions such as *Revolutions—Forms that Turn*, at the 2008 Sydney Biennale, approached avant-garde kineticism predominantly as an historical, machinic tendency that signified the modern aesthetic. While contemporary media art historians such as Edward Shanken, Charlie Gere, Peter Weibel and Ernest Edmonds have acknowledged the works and discussions of avant-garde artists working with kinesis, this acknowledgement is primarily to strengthen the historical genealogy of digital art, where kinetic sculpture is a long lost ancestor (Weibel 2007, pp. 19–41).

This chapter will begin by first unpacking the key understandings of 'Systems Esthetics', which is considered as an attempt to predict the future intersections between art, science, and technology. Secondly, the chapter will highlight Burnham's exclusion of kineticism from 'Systems Esthetics', 'Real Time Systems', and *Beyond Modern Sculpture*, and argue that his position is dependent on works that use movement to explore conceptual and post-formalist objectives. Burnham's emphasis on automata, light art, robotics, and cybernetic art performs a reterritorialisation that moves from kinetic movement towards the movement of systematic processes. This is most prominently addressed through Hans Haacke's early systems art that was made during the 1960s and 1970s, and whose works I address as dependent on actual movement to signify, perform, and process biological, political, and natural weather systems. Burnham's emphasis on the movement of systems art, rather than the affect of kinesis, is more than a syntactical argument. As I will argue, Burnham sequesters the theory and practice of movement in art away from the post-modern aesthetics that were emerging at the time. The effects of Burnham's argument and the popularity of his scholarship in contemporary media arts has reified a regard for kinetic sculpture and installation as a modern mechanical and antiquated practice in art, and is therefore a central influence on contemporary kinetic art history.

4.1 Key Understandings of Jack Burnham's 'Systems Esthetics'

During the 1960s, the term 'systems' was used in a wide range of disciplines that called for an open theory of organisation and communication within scientific, biological and cultural analysis. Many of the systems discourses, analysis, and aesthetics were largely influenced by Ludwig von Bertalanffy's *General Systems Theory*, which regarded biological processes of evolution and adaptation as a number of intersecting systems. Outside the sciences, systems theory was appropriated to understand the flow of information in and across technological media in communications, and it was useful as a model for understanding patterns and processes within economics, chemistry, biology, engineering, sociology, physics, and art as expressed in *Steps to an Ecology of Mind* (1972). Even though there were many interpretations and divergences from Bertalanffy's original theory, systems in general quickly became a framework for understanding how modern society was organised and marked a 'shift in twentieth-century thought toward a biological model' (Kwinter 2001, p. 9) from WW1 up to the late 1960s (Shanken 2009, p. 1).

While the term 'system' is rife in contemporary society, it can also be quite an ambiguous term. A system might be thought of as a number of variables that have the capacity to relate to one another, and which form a larger, rationalised whole (Halsall 2008, pp. 22–34). What is unique about a system is not so much its actual components but the way they are organised. Unlike chaos, systems are defined by the relationships between variables, each of which contributes to the unique form of the entire system. Systems can be quite open—for instance, the entire world can be considered as a total system (Boulding 1985, p. 9)—but there is a key set of criteria that defines them. Systems must have one or more key identifiable functions, and each individual variable within the system must also genuinely contribute to that system's function. If one component is removed from the system, its function is consequently altered from its previous state. From this basic criteria systems can be organised to take in a wide variety of structures, patterns, rhythms, or networks.

Burnham's use of systems theory hinged on the interdisciplinary nature and applicability of Bertalanffy's general systems theory. If a system was regarded as 'a complex of components in interaction' (Bertalanffy 1967, p. 69) Burnham considered art as a system that intersected with all areas of life, from which new subsystems would emerge. As he explained:

> A systems viewpoint is focused on the creation of stable, on-going relationships between organic and non-organic systems, be these neighbourhoods, industrial complexes, farms, transportation systems, information centers, recreation centers, or any other matrixes of human activity. All living situations must be treated in the context of a systems hierarchy of values (Burnham 1980, 'Systems Esthetics' p. 16).

Burnham focused on a system as an open, porous and intersecting structure that encompasses the behaviours, actions and tendencies within all artistic practices that relate to and affect society.

Influenced by Lucy Lippard's reflection on the dematerialised object in the 1960s, and Norbert Wiener's cybernetic theory, Burnham depended on an anti-ontological focus on systems theory in art. As he explained, '[t]he object denoted sculpture in its traditional physical form, whereas the system (an interacting assembly of varying complexity) is the means by which sculpture gradually departs from its object state and assumes some measure of lifelike activity' (Burnham 1968a, p. 10). By diminishing the distinction between art and life through systems theory, Burnham's perspective was a direct response and resistance to Michael Fried's criticism of the theatricality of conceptual and minimalist art. Therefore it was important that Burnham used the term 'Systems Esthetics' rather than 'conceptual art', because the former encompasses the technological expansion and emerging interdisciplinary nature of art in the 1960s—which, for Burnham, key critics such as Michael Fried had undermined.

Similarly to Eco's notion of the 'open work', Burnham was attentive to the prospect of artists such as Hans Haacke considering movement as an open system. For instance in 'Real Time Systems' Burnham discussed the power of actual movement as a means for destabilising the distinctions between art and life (pp. 27–38). The orchestration of actual movement in art became central for achieving this, particularly in a decade of accelerated technological transformation. For Burnham, the real-time affects of actual movement prevent art from being centred on specific objects, and are created to foreground perception as a systematic process of relations between variables. Real-time systems in art therefore easily destabilise the 'ideal time' of modern classical forms of art and reflect how information is being processed in the technological age of the computer.

In 'Systems Esthetics' Burnham considers even a single artwork as its own organised system that resides as a component within the wider system of the art world. Like Popper's focus on demateriality as a relational and communicative tool for artists, Burnham's approach to systems in art is centred on the felt but unseen entities of art objects; where '[t]he specific function of modern didactic art has been to show that art does not reside in material entities but in relations between people and the components of their environment' (Burnham 1968b, p. 16). Following von Bertalanffy Burnham suggests an approach to art that moves away from a focus on the object and towards art as a component within the larger system of society. In doing so, he abandons an inclination towards media specificity because his systems are defined by a 'conceptual focus rather than material limits' (Burnham 1968b, p. 17).

As with Popper's focus of demateriality in participatory art, which I will address in the following chapter, Burnham's systems theory is an articulation of the relationships among viewers, and between viewers and the art. Some of these relationships are engaged through real-time interactions; however, Burnham also specifies that his take on systems theory can be applied to more than the time-based, ephemeral staged environments and happenings. He argued that systems theory, 'deals in a revolutionary fashion with the larger problem of boundary concepts. From a systems perspective there are no contrived confines such as the theater proscenium or picture frame. Conceptual focus rather than material limits define the systems' (Burnham 1968b, p. 7). Because of this post-formalist approach, systems aesthetics is focused; on an expansive and non-representational approach to art that has the potential to be applied to interpretations of art across a variety of practices.

Burnham was working within an array of discussions that focused on the relationships between art and technology, and their future progressions taking place in the 1960s. As David Mellor reflects, the time was marked by excitement around art and technology:

> A dream of technical control and of instant information conveyed at unthought-of velocities haunted Sixties culture. The wired, electronic outlines of a cybernetic society became apparent to the visual imagination. [...] It was a technologically utopian structure of feeling, positivistic and "scientistic" (p. 107).

Hence Burnham's *Beyond Modern Sculpture* optimistically predicted that artists would come to explore new means of visualisation through technological expansion (pp. 185–378).

As a key quintessential exploration of systems theory in art, Burnham's argument drew upon German artist Hans Haacke, and his physiological and biological installations that involved studies of ecological and biological movement such as *Sky Line* (1967), where hundreds of helium-filled balloons were connected together and cast out into the sky from Central Park. Another work, *Photo-Electric Viewer Programmed Coordinate System* (1968b), involved a series of photoelectric sensors installed in the gallery walls with infrared beams fitted at eye level. When viewers entered the space, the light bulbs become active, but light up selectively and in response to the viewer's movement. It is a work that acts as a responsive environment by corresponding to the viewer's movement, while also performing random patterns of action, and confusing the viewer and their interaction with the work. If there is more than one viewer in the space, the system will mimetically light up when two or more viewers move as one body through improvised choreography. As Haacke explains that the work is:

> A 'sculpture' that physically reacts to its environment is no longer to be regarded as an object. The range of outside factors affecting it, as well as its own radius of action, reach beyond the space it materially occupies. It thus merges with the environment in a relationship that is better understood as a 'system' of interdependent processes. […] A system is not imagined, it is real (Haacke in Burnham 1968b, p. 22).

Like Moholy-Nagy's *Light Space Modulator*, a work that I address in Chap. 5 of this book, Haacke explicitly informs his audiences that the *Photo-Electric Viewer Programmed Coordinate System* forms an environment rather than acts as an autonomous object. Both Moholy-Nagy and Haacke develop kinetic motion to cast attention outwards into the space of the gallery. While Moholy-Nagy was more interested in the modulation of movement, Haacke's responsive system is an experimentation of the translation, process, and mimesis of human movement. Despite these differences, there is a continued regard for kinesis as a process that is carried rather than created by material form. From this perspective, Moholy-Nagy and Haacke orchestrate kinesis in their artworks for its material, immaterial, and immanent qualities rather than as a study of purely ontological motion.

However for Burnham, Haacke's work signifies a shift in art that moves away from modern industrial kinesis and towards increasingly intelligent technological systems—a tendency that Burnham predicted to be an inevitable outcome for sculpture after the 1960s (Burnham 1968a, pp. 1–16).

Systems aesthetics is used by Burnham as a key catalyst to regard kinetic art as an antiquated technological art that falls short of the emerging tendencies of conceptual, cybernetic, and robotic art critiques, and responds to the emerging technological age. To make this explicit, Burnham also drew on a range of artists working alongside Haacke to demonstrate that there was an emerging systems aesthetic in conceptual art in the 1960s. This group includes Dan Flavin's fluorescent installations, the minimalist sculptural works by Robert Morris, Les Levine, and Donald Judd. In many cases Burnham draws from systems and cybernetic theory to deconstruct the works. In 'Systems Esthetics' Judd is compared to a computer programmer, while Carl Andre is described as having created assemblages of 'modular forms' (p. 24). Burnham emphasised that Judd and Andre, along with the above-mentioned artists, reveal an ongoing 'technological endeavor' (p. 24) that attempted to intersect the relationships between conceptual and technological experimentations in art.

Despite mention of Flavin, Morris, Levine, and Judd, Haacke's artwork was central to Burnham's understanding of systems-based art. This was openly admitted by Burnham, who stated that '[a]s a close friend of Hans Haacke since 1962, I observed how the idea of allowing his "systems" to take root in the real world began to fascinate him, more and more, almost to a point of obsession'(Siegel 1971, p. 18). Haacke reciprocated with equal appreciation by saying that Burnham had introduced him to systems analysis and was among the first to apply general systems theory to visual art (Bird et al. 2004, p. 102).

Although Burnham's approach to systems theory was distinctly unique because it was the first to apply it to art, his argument also drew from concurrent communication theory, which was popularising Norbert Wiener's cybernetics, and which also focused on the processes of communication. Through Wiener, cybernetic theory became a way for negotiating the dematerialised work of art in computer, electronic, and media art practices, by offering metaphors of software and hardware to describe the relationship between the concept and object in a work of art. Take, for example, Roy Ascott's description of cybernetic art: '[w]hen art is a form of behaviour, software predominates over hardware in the creative sphere. Process replaces product in importance, just as system supersedes structure' (cited in Shanken *Digital Arts and Culture*). For Ascott, cybernetic and systems theory were useful not only for describing the conceptual (software) codes that are signified by the art object (hardware) but also for engaging with the conceptual grounds within art.

The popularity of systems theory occurred concurrently with a number of complementary emerging theories in art theory and criticism. For instance, the dematerialised post-object aesthetics that were popularised by conceptual artists in the 1960s, such as systems theory, concentrated on the construction and organisation of concepts, which in the case of art are signified by their material form. For George Dickie in *Art and the Aesthetic: An Institutional Analysis*, this regard for the ontological nature of art emphasised the way in which objects were organised and classified, rather than the form of their physical properties. Therefore, art was no longer considered an emotive expression of the artist. Instead, artists increasingly worked within a system of observation and differentiation from other artworks and contexts (Dickie 1974, pp. 50–52).

Like Dickie, Arthur Danto has likened the artworld of the 1960s to an entire regulated system (Danto 1977, pp. 571–584). Within this artworld artists referred to art historical tendencies to inform their practice, and their emphasis on the conceptual properties of a work of art contributed to Danto's description of the art world as a 'style matrix' that is built and organised by artists and institutions. Danto's perspective was largely informed by his reaction to the Minimalist and Pop Art tendencies that were emerging in art at the time, both of which played conceptual games with the institutional expectations and definitions of art that determined how art appeared and was constructed in exhibition environments. Danto's systematic approach to art theory, history, and criticism reified a definition that was determined not by form or expression but, a classification and organisation of ontological and conceptual objects that formed the network of the system.

In *Art as a Social System* Niklas Luhmann (2000) has since reflected that for many, art has become a social sub-system for society. Likewise, rather than approaching a work of art as an autonomous object from which meaning can be drawn, Burnham used systems theory to consider art as something that is built from a matrix of components that were organised by artists and institutions. This point of view focuses on the communicative relations that exist between the work, its viewers, and the historical context rather than the components themselves.

Norbert Wiener's theory of cybernetics similarly emphasised the connections among variables in technological information networks. For Wiener, the 'second industrial revolution', kicked off during domestication of computing technologies and placed an emphasis on the input and feedback of information systems. Like systems and post-object aesthetics, Wiener viewed the information systems of computing technology as being guided by processes, rather than objects—or as Ross Ashby has said, cybernetic theory emphasises 'not things, but ways of behaving' (p. 1). And, as art historian Charlie Gere, has stated, Wiener's cybernetic theory was a rubric for approaching 'biological, machinic and social processes' across a number of disciplines outside the sciences, including art' (Gere 2002, p. 52). Although Wiener and Ashby were reflecting on information systems prior to the 1960s, cybernetic and systems theory intersected with one another most prominently during this time. A key example was Burnham's 'Sculpture as System', in *Beyond Modern Sculpture*, which will be addressed later in this chapter. But first Burnham's critique of kineticism in art must be addressed in order to understand how and why the 'kinetic kraze' quickly shifted to become an unrequited forerunner of technological art.

4.2 Burnham's Turbulent Relationship with Kineticism

As Pamela Lee (2004) has warned in *Chronophobia*, the problem with constructing a system, even one that is thought of as being open, is that there always outliers to the system (pp. 243–246). Systems theory can be problematic when using it as a potential dominant mode of thought in art history, theory, and criticism, as Burnham has because the system can easily become a ridged formula for omission from the arts. Just as Lee has warned, the exclusion of the study and orchestration of actual movement in art from contemporary criticism in some ways locks kinetic artworks into a zone of anachronism, and disassociated them from the contemporary.

Burnham's focus on systems theory in the 1960s was a catalyst for the turbulent relationship Burnham had with the art, theory, and criticism of kinetic sculpture. I say turbulent because despite kinetic sculpture was considered to be one part of technological art that Burnham described as the 'panacea that failed' (Burnham 1980). Additionally despite Burnham's exclusion of kinesis in 'System Esthetics' and *Beyond Modern Sculpture*, systems theory has played an important role for many kinetic artists in contemporary art history. For Haacke for example, actual movement in art has been used as an effective tool for rendering visible the unseen relationships among variables. This means that movement is not necessarily used in a formalist sense by exploring the rhythms of movement on a purely visual basis but can be a means for presenting how relations between components move in systems.

Burnham's view on the emerging experiments with art and technology was also inconsistent. A few short years prior to the publication of 'System Esthetics',

Burnham wrote damning reviews on the collaboration between artists and engineers in a series of exhibitions such at the Los Angeles County Museum of Art's (LACMA) exhibition *Art and Technology* (1971). In a review for the exhibition in *Artforum* Burnham wrote:

> If presented five years ago, A&T would have been difficult to refute as an important event, posing some hard questions about the future of art. Given the effects of a Republican recession…few people are going to be seduced by three months of industry-sponsored art, no matter how laudable the initial motivation (1980, p. 210).

Burnham's frustrations with other exhibitions and collectives such as Experiments in Art and Technology (E.A.T), *Cybernetic Serendipity*, at the Institute of Contemporary Arts in London (1968b), the Center for Advanced Visual Studies at the Massachusetts Institute of Technology, and his own exhibition *Software*, at the Jewish Museum in New York (1969), are detailed in his essay 'Art and Technology: The Panacea that Failed'. Burnham regarded the above exhibitions as run by elitist organisations with compromised funding from corporate sponsors and inadequate financial support from artistic and educational institutions (pp. 200–202). In the same essay Burnham also criticised these institutions and the artists in the exhibitions for showing an overall technological incompetence, and for failing to be experimental with cutting edge technologies in art (pp. 211–215).

Additionally, Burnham's curatorial role for the exhibition *Software* (1970), experienced a considerable amount of public controversy, and several artists threatened to withdraw from the exhibition. Burnham claimed that artists had been sabotaging their own and other artists' works and said: 'the results have fared from mediocre to disastrous when artists have tried to use what has euphemistically been referred to as the electronic technology of "postindustrial culture" and lacked an aesthetic competency' (Burnham 'Art and Technology' p. 200). These factors stood out for Burnham as reasons why artists were so far unable to create 'socially acceptable art' that utilised the latest emerging technologies and theories related to art at the time (p. 200).

Burnham's criticism towards those collaborating with science and technology in their art at the time, were often directed at artists experimenting with kinesis. The popularity of kinetic art in the 1960s became, for Burnham, a key reason for why technological art was not being interpreted through general systems theory. In his words:

> By the fact that most systems move or are in some way dynamic, kinetic art should be one of the more radical alternatives to the prevailing formalist esthetic. Yet this has hardly been the case. The best publicised kinetic sculpture is mainly a modification of static formalist sculpture composition. In most instances these have only the added bonus of motion, as in the case of Tinguely, Calder, Bury, and Rickey….All too often gallery kinetic art has trivialized the more graspable aspect of motion: this is motion internalized and experienced kinesthetically (Burnham 1968b, p. 22).

For Burnham, mechanical kinesis was at the time too closely connected with formalist tendencies, and this resulted in a continued desire to create motions purely from physical systems.

As a former lumia and kinetic artist, Burnham's critique is largely cast by his disappointment with artists working with kinesis to resist the emerging popular anti-modernist perspectives on art. As he explained, '[t]he important thing is that the Kineticist is trying to make himself relevant in a world which is continually being recreated', and from which they were falling behind (Burnham 1968a, p. 284). Therefore, Burnham's description of kinetic art as an unrequited practice was made in dissatisfaction with electro-mechanical sculpture, because artists had the unrealised potential to assert kinesis within the emerging postmodern perspectives. As he indicates in *Beyond Modern Sculpture*:

> In an art world of 'cool' stances and exploding values a dialectical tension continues to build around Kineticism. The very fact that it is unrealized art should remain a sign of encouragement for future artists, even though Philip Leider suggests that Kinetic artists should enlarge upon the stance of Tinguely and create more self-destroying machines. Actually, even with the desire attached to that death wish, it is, at best, only a Luddite solution. The real way to kill an art movement is for it to realize its goals—an objective which Kineticism has yet to achieve' (Burnham 1968a, p. 284).

Therefore, Burnham's criticism of kineticism in art, as a practice that reflected the social and technological issues that were contemporary at the time, was also about the unrealised potential for artists to use kinesis in a way to re-engage with society and once again produce a dominant form of technological media art.

Despite Burnham's adamant exclusion of kineticism from 'Systems Esthetics', in the same year that Burnham published the essay in *Artforum* Willoughby Sharp referred to kinetic artists as the forerunners of exploring systems theory in art. Sharp said that systems 'are defined by their energy input.... They are a cohesive collection of components relating to a single set of systems equations. These systems deal with facts about our physical reality.... One of the major functions of these sculptural systems is to plug us into the actual forces that configure contemporary reality' (p. 10). For Sharp, kinetic art not has the capacity to perform the rhythms of movement and energy that flow through daily life but, also act as a manifestation of how reality is perceived and framed by society in different ways.

Sharp framed kinetic sculptural systems as a truly avant-garde practice that breaks down the boundaries between art and life, where:'[p]ainting and static sculpture are obsolete. They no longer relate to reality. They are anachronisms because they are irrelevant to our contemporary technological situation. It's idiotic and immoral to make such objects as art now' (p. 4). Therefore, for Sharp, whether kinetic art is made from mechanical, biological, pneumatic, or electronic media, it has the capacity to perform visual manifestations of life as a series of systems.

Despite his disappointment in technological experiments in art during the 1960s and 1970s, it cannot be ignored that a number of works by kinetic artists also strongly influenced Burnham's perspectives on the use of general systems theory in art. This suggests that Burnham's critical shift away from kinetic movement and towards the movement of systems was in part a syntactical maneuver to disassociate it from a connection to modern art. Consequently, Burnham attempted to remove movement entirely from postmodern interpretation of art. He argued that artists working with new technological media at the time were the forerunners of systems theory in art and were developing new ways for visualising and conceptualising a systems approach to art in real-time. According to Burnham, this tendency was informed by early European avant-garde constructivist and productivist collectives and artists who drew from the modern industrial machine aesthetic, such as Moholy-Nagy, who was considered to have pioneered a systematic approach to his sculpture, photography, and painting. As Burnham explains:

> In his book, *The New Vision*, Moholy-Nagy described fabricating a set of enamel on metal paintings. These were executed by telephoning precise instructions to a manufacturer. An elaboration of this was projected recently by the director of the Museum of Contemporary Art in Chicago, Jan van der Marck, in a tentative exhibition, 'Art by Telephone'. In this instance the recorded conversation between artist and manufacturer was to *become part of the displayed work of art*. For systems, information, in whatever form conveyed, becomes a viable esthetic consideration [author's italics] (Burnham 'Systems Esthetics' p. 28).

Other kinetic artists were crucial to Burnham's formulation of an aesthetic of systems in art. Among these are works by Len Lye, Otto Piene, and Robert Breer's floats, and Group de Recherches d'Art Visue (GRAV) (a kinetic art collective that included Julio le Parc, François Morellet and Yvaral).

It is in *Beyond Modern Sculpture* that Burnham's criticism of kineticism is addressed in further detail, he devotes an entire chapter to his argument that artists working with kinesis are inevitably unrequited by their art. Burnham's main intention for systems theory in art was not only to develop a way of understanding the emerging forms of contemporary art in an increasingly technological society but, also that art as a system would become the dominant framework for creating, experiencing, and distributing art in the future. This claim was explicit in his lecture at the Guggenheim museum in 1969, 'The Aesthetics of Intelligent Systems' when he said:

> Although the art of the future could take any one of a number of directions, it seems to me that, with the steady evolution of information processing techniques in our society, an increasing amount of thought will be given to the aesthetic relationship between ourselves and our computer environments—whether or not this relationship falls into the scope of fine arts' (quoted in Gere 2002, p. 129).

Through Haacke, Burnham considered that movement processes were an effective way for communicating systems theory in art, to the point that 'real-time information processing mode [was] rapidly becoming the routine style of handling information' (Burnham 1969, p. 30). Burnham continues: '[w]hat a few artists are beginning to give the public is real-time information, information with no hardware value, but with software significance for effecting awareness of events in the present' (30). This connection between concept as software and material as hardware compares art to processing systems and excludes an interpretation and discussion of the way these systems *move*. Burnham's prediction that systems aesthetics would become a dominant approach to art in an increasingly socio-technological context is based on the dismissal of movement as a means to signify, process, and perform systems art.

4.3 Hans Haacke's Kinetic Systems

Burnham, however, failed to acknowledge that Haacke's early systems works performed kinetic dynamism to express systems processes. This failure is problematic especially when contemporary artists and scholars such as Ernest Edmonds has highlighted actual movement, dynamism, or change are key characteristics of systems in art where an art object "consists of a system that changes within itself and where that change is apparent to an observer" (19) and which "an art object may or may not be produced" (Cornock and Edmonds 1973, p. 11). Here physically noticeable movement is an integral aspect to systems in art regardless of whether the artwork is creating or reacting to change. Furthermore 'interactive art systems' might entail smaller unnoticeable movements that might still influence the movement or appearance of an artwork" (20).

Rather than discussing Haacke's use of movement to signify and perform information systems, Burnham instead described Haacke's work with a vocabulary that spatialises the temporal movements of kinetic dynamism. Luke Skrebowski has recently argued that these early sculptures emphasised the movement of 'physiological, physical and biological processes' and are central to Haacke's application of general systems theory, which have persisted throughout his artistic career (2008, pp. 59, 77). In his analysis, Skrebowski problematises Benjamin Buchloh's attempt to create a division between the artist's biological kinetic works and his more politically engaged art. He argues that to draw such a distinction is a reductive binarism that patronises the complexity of Haacke's early practice.

Although Skrebowski does not address this, his emphasis on Haacke's early sculptural works also problematises Burnham's antipathy towards the intersection between kinetic dynamism and systems art. In his attempt to exempt kinesis from post-modern interpretation, Burnham's argument overlooks Haacke's emphasis on ontological function and form as referents for conceptual systems in art. Rather, for Haacke, it was essential that the real-time processes and conceptual systems were signified in his art, performed on a material level. As Haacke said, 'I was primarily what you might call job-oriented. Even in the '60s, I wanted things to function, in a very literal, physical sense' (in Buchloh 'Hans Haacke' 1988, p. 220). It is these material kinetic systems that I would like to bring attention to, rather than Haacke's later socio-political systems art. Not only are Haacke's early kinetic installations from the 1960s important features of contemporary kinetic and new media art history but his emphasis on the movement of form as a referent for material and conceptual systems is complementary to the view of kinetic art throughout this book.

For instance, early installations such as Haacke's, *Blue Sail* (1964–65), *Condensation Cube* (1963–65), and *Sky Line* (1967) have often been dismissed as experiments in 'positivistic scientivism' (Buchloh 1988, p. 212), all too preoccupied with technological rather than conceptual experimentation. Haacke's systems art is more in line with Edmonds' approach to 'interactive art systems' which is open to the physical and informational movements and interactions between an artwork and its audience (20). For instance works such as *MoMA Poll News* (1969), *Shapolsky* et al. *Manhattan Real Estate Holdings, a Real-Time Social System, as of May 1971* (1971), which are works that each collate and display data over the duration of the exhibition. Each work materialises the invisible systems that traverse the everyday, such as a constant influx of political news, the hierarchical structure of real estate in New York, or the demographics of attendees at Haacke's exhibitions.

Yet much of the systems art by Haacke in the 1960s and early 1970s was created with an emphasis on the unstable nature of material through time-based movement and transition. For example, *Condensation Cube* (1963–65), *Ice Stick*, (1966), *Ice Table* (1967), and *High Voltage Discharge Traveling* (1968) are all sculptures in which Haacke prioritises the movement of natural-process energy systems like condensation, precipitation, evaporation, and the expansion and contraction in temperature change. Haacke also explored the movement of evolution, reproduction, birth, and death with *Grass Cube* (1967), *Live Airborne System, November 30, 1968* (1968), *Grass Grows* (1969), *Chickens Hatching* (1969), *Transplanted Moss Supported in an Artificial Climate* (1970), *Bowery Seeds* (1970), *Goat Feeding in Woods* (1970), *Directed Growth* (1970–72), and *Rhine Water Purification Plant* (1972). As with Haacke's data-processing systems art addressed earlier, all works listed here are described as 'event containers' that render visible the unseen movements of natural elements in the gallery setting (Grasskamp 36).

Haacke's use of movement also continues many of the experimentations with movement by early European avant-garde artists that are often recalled in contemporary art history. Like artists before him, such as Moholy-Nagy and Gabo, Haacke was concerned with rendering visible the ordinarily invisible facets of motion and energy. While Moholy-Nagy and Gabo were concerned with space-time perceptions, or what Moholy-Nagy termed the 'dynamic construction system of forces' (Moholy-Nagy 1965, p. 238), Haacke was interested in collapsing the distinction between life and art by performing actual motions in life, including the life cycle, metabolism, and the transfer of energy.

Take, for instance, Haacke's *Chickens Hatching* (1969), an installation that consists of fertilised chicken eggs, incubators, a lamp, and a thermostat, which is both an exploration of kinetic movement and biological systems in art. For Burnham, *Chickens Hatching* presents a system 'where information is derived from the normal activities of animals in their environments'(Burnham 'Real Time systems' 30). This is an example of 'real-time information, information with no hardware value, but with software significance for effecting awareness of events in the present' (28). While there is little specific hardware value to the actual installation in that time and place, the actual, emergent, and material movements and behaviours of the chickens are the material processes, which unlock the conceptual systems within the work.

It is important to note that these kinetic systems works were created throughout the 1960s and early 1970s during the same time that Haacke was creating works that processed the information of social and political systems. Haacke's early artistic career did not move away from kinetic dynamism towards systems art in a clean transition; rather, kineticism was used early on to perform and experiment with the natural, biological, and technological transformation of energy. This understanding renews a connection between kinetic and conceptual art and, as I have suggested, also problematises Burnham's regard for kineticism as a practice that is inherently connected to the modern industrial revolution.

To quote Haacke, these works 'make something which experiences, reacts to its environment, changes, is non-stable…something which the "spectator" handles, with which he plays and thus animates…something which lived in time and makes the "spectator" experience time' (in Brett 2000, p. 294). Haacke's emphasis on the subjective perception of time resonates with Bergson's study on the perception of duration, as a means of experiencing time, while concentrating on duration as a process of constant change. As temporal-pieces, Haacke therefore presents constructions of 'natural' time in the sterile environments of artistic institutions to present society in an age of technological expansion that has, in a sense, lost time, despite its fixation on efficiency.

While Haacke has since been connected with systems aesthetics, biological art, and process art when considering the nexus between time and kinesis, the movements that Haacke orchestrates portrays time as a system of durations. Although there were moments when Haacke resisted the term 'kinetic' to describe his works (Jones 2011, p. 9), the use of movement within many of his works during the 1960s was a focal aspect to his practice. Kinesis, for Haacke, was more closely related to entertainment, rather than art, and kineticism became a term synonymous with public amusement rather than a means for widening and reflecting on the conception of art. His elaboration on his sculptures as 'time systems' in the 1960s resonates with Umberto Eco's approach to kinesis as an example of 'open works'. For Haacke, they 'merge with the environment in a relationship that is better understood as a "system" of interdependent processes' (quoted in Burnham "Systems Esthetics", p. 35). To isolate the movement of a body of water into a perspex container works exactly to demonstrate that movement , even when it is enclosed, refers to and affects other movements (in this case the movements of Haacke's spectators interacting with the piece).

Haacke's *Blue Sail* (1964–65) is a simple motion study consisting of a sheet of blue chiffon, approximately 3.4 m long and 3.2 m wide, suspended horizontally in the air from the ceiling, and weighed down with fishing weights. Underneath the blue sail stands a small domestic fan pointed up towards the sail and panning across it. The sail is porous enough to form the shape of the wind created by the fan without entirely billowing and rising upward. The movement of the fan is translated by the sail; it creates a wave that hovers in constant equilibrium.

This work is one of Haacke's closed environmental 'sculptural' systems works from the early 1960s. His focus on creating sculptures that produce their own weather systems, like other participatory art at the time, highlighted the tangibility of art and what Fried would call the duration of its objecthood. The *Blue Sail* presents a movement that is perpetual. It moves in front of its viewers, as well as when it is alone in the exhibition space. As Haacke explains:

> A 'sculpture' that physically reacts to its environment is no longer to be regarded as an object. The range of outside factors affecting it, as well as its own radius of action, reach beyond the space it materially occupies. It thus merges with the environment in a relationship that is better understood as a 'system' of interdependent processes. These processes evolve without the viewer's empathy. He becomes a witness. A system is not imagined, it is real (cited in Burnham 1968b, p. 22.).

Movement and time are important elements of Haacke's exploration of systems theory in art. Motion is orchestrated to resist a modern definition of art as autonomous, finite, and dependent on its crafted form, by producing work like *Blue Sail* to alter and work within the interior climate of the gallery space and function in time. The components of *Blue Sail* each respond to one another where the movement of one object (a fan) causes an effect of movement on another (the sail), which also interacts with, and is affected by, the space and the viewers within it. The unstable or sensitive relationship that *Blue Sail* has with its environment emphasises movement as an unfolding process that is not contained but open to its environment. It is a work that, like many of Haacke's early systems, 'evolve[s] in time and [is] affected by time'; it is persuaded by an objective temporality rather than the phenomenological 'shifting experience of the viewer' (Haacke).

Blue Sail is a work that Haacke created in order to make something which performs in time and makes the viewer bring attention to the experience of time. On another level, unlike many participatory and interactive artworks at the time, its movement performs with independence from its viewers; it moves in time and also affects time in Haacke's attempt to emphasise the transformation and process of energy movement systems.

It is important that works such as *Blue Sail* and *Photoelectric Viewer-Controlled Coordinate System* perform Haacke's approach to systems theory, rather than create a database or visualisation of them, unlike Haacke's socio-political systems such as those written about in *News*, *MOMA Poll*, and *Gallery-goers' Birthplace and Residence Profile*. Providing a visualisation of these systems would create an abstraction that distorts the unfolding nature of temporality. Kinesis is, therefore, a central tool for illuminating and performing the specific temporal arrangements in Haacke's systems art.

4.4 Conclusion: Ontologically Unstable Movement Systems

Burnham's de-emphasis of movement in Haacke's early systems art was used to defend his arguments made in 'Systems Esthetics', 'Real Time Systems' and *Beyond Modern Sculpture*. In these texts Burnham argues that art in the late 1960s was increasingly moving away from an orientation of objects and towards a systems-based approach to creating and consuming art. The consequence, as Burnham argued, is that it positions kinetic artists as unrequited in their aims to contribute a popular technological arts practice. The popularity and influence of

Burnham's argument in media art history has contributed to an understanding of kinetic sculpture as a practice solely associated with the industrial machine aesthetic and modern avant-garde movements. And yet contemporary scholars such as Edmonds who understand movement to be a key defining characteristic of systems theory in art signify that Burnham clearly overlooked the capacity for kineticism to participate in experiments between art and technology at the time.

This chapter has also highlighted the works and writings of Hans Haacke, who in the 1960s and 1970s drew from Norbert Wiener's cybernetic theory and Ludwig von Bertalanffy's systems theory to create works that perform, signify, and unfold as biological, political, and natural systems to challenge the institutional boundaries between art and life. I have argued that Burnham's interpretation of kineticism misdirects Haacke's early sculptural systems art as early information-processing systems, rather than experiments of movement systems. Contrary to Burnham, I have argued that a sensitivity for, and discussion of, the actual kinetic movement in Haacke's works is a central aspect of the artist's understanding of systems aesthetics in art. Therefore, while Burnham's antipathy towards kineticism separated kinesis from the emerging systems aesthetics in art, artists such as Haacke explicitly emphasised the importance of movement and form to connect media with specific conceptual messages. Haacke created works that were ontologically unstable in order to highlight the unfolding entropic approach to time that is found within systems theory. Rather than deferring to actual movement, Haacke used motion to highlight it as a tool that is material and immaterial, as well as actual and virtual. This approach to kineticism resonates with the use of movement by many of the other artists who are referred to in this book: Moholy-Nagy, Gabo, Tinguely, and Kapoor.

Burnham's 'Systems Esthetics' is also significant for considering how kinetic art has since been interpreted as a technological precursor to sculptural and installation media art that emerged in the 1970s. The following chapter approaches the ways in which conceptual, participatory, and communicative time-based arts have been considered as dematerialising the form of kinetic sculpture to the point of invisibility and further associate kineticism with an antiquated modern machine aesthetic.

References

Anonymous (1966) Styles: the movement movement. Time 87:66–69
Ashby R (1957) An introduction to cybernetics. Chapman Hall, London
Bertalanffy L (1967) Robots, men and minds. G. Braziller, New York
Bird J, Grasskemp W, Nesbit M (eds) (2004) Hans haacke. Phaidon, London
Boulding K (1985) The world as a total system. Sage, Beverly Hills, California
Brett G (2000) Force fields phases of the kinetic. Hayward Gallery, London
Buchloh B (1988) Hans Haacke: memory and instrumental reason. Art in America 76(2):203–241
Burnham J (1968a) Beyond modern sculpture: the effects of science and technology on the sculpture of this century. G. Braziller, New York
Burnham J (1968b) Systems esthetics. Artforum 7(1):30–35

Burnham J (1969) Real time systems. Artforum 8(1):49–55
Burnham J (1980) Art and technology: the Panacea that failed. In: Woodward K (ed) The myths of technology; technology and postindustrial culture. Routledge, London
Cornock S, Edmonds E (1973) The creative process where the artist is amplified or superseded by the computer. Leonardo 6:11–16
Danto A (1977) The artworld. In: Madoff S (ed) Pop art: a critical anthology. University of California Press, Berkeley
Dickie G (1974) Art and the aesthetic: an institutional analysis. Cornell University Press, Ithica
Edmonds E (2011) Interactive: art, research and the creative practitioner. Libri Publishing, Faringdon
Gere C (2002) Digital culture. Reaktion Books, London
Halsall F (2008) Systems of art; art, history and systems theory. Peter Lang, Oxford
Jones C (2011) Hans Haacke 1967. MIT List Visual Arts Center, Cambridge, Massachusetts
Kwinter S (2001) Architectures of time, toward a theory of the event in modernist culture. MIT Press, Cambridge
Lee P (2004) Chronophobia: on time in the art of the 1960s. The MIT Press, Cambridge
Luhmann N (2000) Art as a social system. Stanford University Press, Stanford California
Mellor D (1993) The sixties art scene in London. Phaidon, London
Moholy-Nagy L (1965) Vision in motion, 2nd edn. P. Theobald, Chicago
Shanken E (2009) Reprogramming systems aesthetics: a strategic historiography. Digital Arts and Culture. Retrieved from: http://escholarship.org/uc/item/6bv363d4
Sharp W (1968) Air art. Kineticism Press, New York
Siegel J (1971) An interview with Hans Haacke. Arts Mag 46(7):18
Skrebowski L (2008) All systems go: rediscovering Hans Haacke's systems Art. Grey Room 30:54–83
Weibel P (2007) It is forbidden not to touch: some remarks on the (Forgotten Parts of The) history of interactivity and virtuality. MIT Press, MediaArtHistories, Cambridge

Chapter 5
Digital Determinism: A Post-kinetic and Dematerialised Time

Abstract This chapter argues that Frank Popper's writings on kinetic sculpture have influenced the assumption that kinetic art is purely a formal investigation into movement, time, technology, and art. This misconception stems from Popper's approach to dematerialty and technological determinism. Popper associates mechanical media with materiality and digital media in art with immaterial and virtual qualities. Popper's perspective reifies the assumption that kineticism is a purely modern investigation of the movement of mechanical media and ignores the immaterial qualities of key modern artworks, as well as the material aspects of artworks with digital media. I argue that rather than considering kinesis as a practice that became increasingly invisible, as Popper has suggested, kinesis can be used to articulate movement as an emergent material process that arrives in the present from a state of preacceleration.

In 1968 Frank Popper published *Origins and Developments of Kinetic Art*, the same year as Jack Burnham's 'Systems Esthetics' and *Beyond Modern Sculpture*. To date, *Origins and Developments of Kinetic Art* is the most comprehensive analysis of kinetic sculpture as an aesthetic art form in Europe and North America up until the late 1960s, and Popper has since been recognised as 'the foremost European historian of art and technology' (Shanken and Ascott 2003, pp. 172, 184).

Since then Popper primarily focused on the progression and future of art in an increasingly technological society. This chapter argues that, as with Burnham's 'Systems Esthetics' and Fried's (1998) 'Art and Objecthood', Popper's scholarship presents a polemic for how kinetic sculpture is framed in art history today. Popper's approach to kineticism in *Art, Action and Participation* and *From Technological to Virtual Art* regards kinetic sculpture as a modern antiquated experiment with mechanical media that is not relevant to contemporary art and theory. His assumption that digital art is 'more refined' (2007, p. 1) than analogue media regards kinetic sculpture as a 'forgotten art' which can only be 'excavated' by art history (Borja-Villel p. 7). Such an assumption is problematic because it steers a

© Springer Nature Singapore Pte Ltd. 2017

C. Chau, *Movement, Time, Technology, and Art*, Springer Series on Cultural Computing, DOI 10.1007/978-981-10-4705-3_5

focus away from discussing the actual and physical movement of hardware involved with digital art in movement, and also omits kineticism from contemporary discussions about movement in art today.

Popper's later work is directed towards a digital and immaterial experience of art that moves. This transition coincides with what he describes as a 'post-kinetic' tendency in technological art since the 1980s (1987, pp. 301–302). Throughout his subsequent publications, *Art, Action, and Participation* (1975), *Art and Electronic Media* (1993), and *From Technological to Virtual Art* (2004), Popper has analysed media art alongside the discourses that surround traditional modes of representation because of its increasing popularity with artists, institutions, and audiences.

In doing so, Popper's scholarship from 1968–2004 forms an historical analysis that moves from modern mechanical kinetic sculpture, through to digital art practices such as virtual art, multimedia online art, and multi-sensorial installation. In the process, kinetic sculpture becomes increasingly regarded as a modern mechanical formal exploration of actual movement. Rather than considering the role and effects of kineticism in contemporary art, Popper assumed that media artists have naturally progressed towards a digital immateriality (1975, p. 278). While Popper contributes to a contemporary understanding that avant-garde kinetic artists have a tradition with modern mechanical media, his position is a technologically determinist account that privileges the technological progress of digital media in art.

Popper's analysis of kinetic and digital art is also significant because his perspective resonates with multiple elements of Terry Smith's popular analysis of contemporary art. Popper's view that contemporary digital media art is the conceptual, immaterial, and 'more refined' technological media nurtures a remodernist understanding that contemporary digital art destabilises previous modes of representation. Additionally Popper's association with kinetic sculpture as a modern mechanical and formal experiment with movement also provides an opening for contemporary exhibitions such as *Revolutions—Forms that Turn* as attempts to collapse chronological distinctions between modern and contemporary art.

This chapter will unpack Popper's interpretation and appropriation of the concept of demateriality, as his explanation for the abandonment of kinetic art, and the popularity of relational and communicative motion in participatory, interactive, and electronic media in the 1970s. I highlight Popper's understanding of demateriality to demonstrate that there is a problematic binary distinction between mechanical media as purely material and digital media as immaterial and virtual. Secondly, the chapter will address Popper's interpretation of digital media that is used to reify an association of digital art as inherently virtual, immaterial, and ephemeral.

Thirdly, the writings by Susan Ballard and Erin Manning are presented as key resources for considering the effects of contemporary kineticism that problematise Popper's perception of kinetic art. Rather than considering kinesis as a practice that became increasingly invisible, as Popper has suggested, kinesis can be used to articulate movement as an emergent material process that arrives in the present from a state of preacceleration. Doing so also emphasises kinesis as a tool for expressing and articulating perceptions of temporality, space, and dimensionality, and diverging from Popper's digital determinism that strictly aligns digital media with immateriality and virtuality, and mechanical media with materiality.

5.1 Popper's Post-kinetic Analysis

To date Popper's *Origins and Developments of Kinetic Art* (1968) is one of the more comprehensive historical analyses of the influences and tendencies of European and North American kinetic art up until the 1960s. Since then, art historians and critics have not surveyed kineticism in art to the same degree of detail, breadth, and awareness. Popper's formal analysis of modern kinetic sculpture explores the use of movement with technology in art between the late nineteenth and mid-twentieth century. Like Jeremy Benthall who predicted that kinetic art would be a practice from which new art forms would grow (Benthall 1972, p. 101), Popper believed that the technological expansion of the 1960s broadened the vocabulary of technological visual media, and provided new ways to present movement (Popper *Art, Action and Participation* 13–51). Consequently according to Popper, kinetic art became a prominent form of technological art after WWII (1968, pp. 121–150).

However, in his publications subsequent to 1968: *Art, Action, and Participation* (1975), *Art of the Electronic Age* (1993), and *From Technological to Virtual Art* (2007) Popper gradually concentrated on other emerging practices such as computer, cybernetic, and digital art. Through this Popper presents a progression that moves away from mechanical kinetic sculpture and installation towards the 'more refined' technological media of virtual art and digital media (Origins and Developments of Kinetic Art 121–150). Popper constructed a lineage that characterised digital media as instantaneous, immaterial, and virtual and treats kinetic art as an early mechanical precursor to contemporary media art.

For Popper, this transition occurred during the 1960s and 1970s when conceptual artists increasingly began to challenge the notion of the art object and approach an artwork as a series of processes and perceptions rather than a discrete object. For Michael Fried (1998), this perspective nurtures a literalist tendency that reduces an artwork to an object within a spectacle of theatricality (pp. 149–172). However, others such as Harold Rosenberg argued that artists during this time approached their art as an, 'arena in which to act rather than as a space in which to reproduce, redesign, analyse or "express: an object, actual or imagined"' (1965, pp. 23–59).

For Rosenberg temporality was increasingly becoming an important tool for exploring and expanding available modes of representation because 'what was to go on the canvas was not a picture, but an event' (p. 23). He emphasised that modern American artists were at the forefront of this way of approaching and defining art, who increasingly considered art as a process rather than a discrete object. Consequently, engaging with the conceptual aspects of an artwork became the focal point of the experience of art, rather than its material elements.

Consistent with Rosenberg, Lucy Lippard later highlighted that artists in the 1960s increasingly regarded the materiality of their works as secondary, or carriers for the ideas that they wanted to communicate. Lippard also broke from Fried's formalist perspective that meaning should be embedded in the formal elements of the art object. By privileging the conceptual components of an artwork, Lippard (1997) argued that for conceptual artists, material form often became 'secondary, lightweight, ephemeral, cheap, unpretentious and/or "dematerialized" in relation to the idea that it signified' (p. vii). This approach destabilised the ontological high and low power structures laden within modern art because of the disregard for medium specificity (pp. 5–11). Instead of focusing on technicality, craft, or formal capabilities, demateriality catalysed a separation between form and concept.

In *Art, Action, and Participation* Popper used demateriality to explain the emerging post-object tendencies of both conceptual and technological artists that Lippard and Rosenberg previously argued. He identifies that a strong selection of artists in the early 1960s and 1970s began to use technological media as a means of expressing their ideas, rather than exploring anything inherent in the technology that they were using. Following Lippard, Popper argued that media artists who were interested in using interaction with technology and/or the dynamics of viewer participation considered the medium of their art as a secondary element to the communicative, relational, and conceptual aspects of the work. Popper's interest was therefore focused on the intersections between conceptual and technological art, as well as the way participation was being used during this time to dematerialise the art object. For Popper this was evident in the works by Fluxus artists, E.A.T, Nouvelle Tendence, GRAV, ZERO, Lygia Clark, Milan Dobes, Allan Kaprow, John Cage, and Robert Rauschenberg, who were amongst the early generation of artists who used participation as a primary aspect of their work.

For example, E.A.T, which was formalised as a collective of artists and engineers after the collaborative exhibition *9 Evenings: Theatre and Engineering* (1966), organised by Robert Rauschenberg and Billy Klüver, consisted of a series of performances and artworks for viewers to interact with over time. One of the more famous works at the exhibition was Rauschenberg's *Open Score*, where a game of tennis was played in the exhibition space. At moment when a player hit the ball a light that was illuminating the game would be extinguished until the players were in complete darkness, while in the meantime a performance with five hundred volunteers was recorded by infra-red cameras and projected onto screens in the space. For Popper, these artists focused on the social interaction with technology to resist traditional modes of producing and experiencing art. Rather than focusing on the movement of form, such as with early kinetic art, (Brett 2000, p. 9–68) artists were exploring the social and perceptual movements that unfolded between viewers and the artwork.

For Popper another key example of participatory art that dematerialises the object of art is Allan Kaprow's happenings, which began in 1959. Kaprow orchestrated a series of participatory performances that have been considered as a synthesis of assemblage, environment, action painting, and intervention, and involved vaguely scripted activities for the artist and his audience to perform together simultaneously. For Popper, Kaprow's happenings enforced an 'anxiety' around an object of art and counts 'on the presence of spectators and sometimes the artist himself- of a situation or event in which elements of everyday life or everyday technology become open to the "strangeness" of the fantastic or the poetic' (1975, p. 22).

Negotiations between art, time, and technology were also discussed amongst artists and critics interested the formal plastic arts in the 1960s. For instance, at the Venice Biennale in 1966 Julio le Parc was awarded the grand prize for painting for his optical and kinetic devices that could be worn and interacted with by viewers. le Parc presented mirrors, sculptures, and wearable objects that altered the viewer's perception to emphasise the experience of art as a process of sensations. le Parc emphasised the movement of participation, interaction, and duration to encourage viewers to reconsider prior conceptions of art as a unique, finite, and discrete object.

With le Parc, the process of engaging with the concept becomes the primary focus, rather than commending the craft of an autonomous work of art. Accordingly, a specific relationship between the form and concept of the work emerges, where the meaning of the work is no longer embedded in its material elements. Here, kinesis is used to rupture and dissolve traditional categorisations between the plastic and temporal arts.

Popper explained that thinking about movement in art that focuses not on kineticism but towards the movement of interaction and communication in participatory art, '[t]he work loses its materiality, and becomes simply an effect or an event' (1975, p. 8). In the process of the event unfolding '[t]he emphasis lies not on the object, but on the dramatic confrontation or the perceptual situation in which the spectator finds himself' (p. 11). For Popper, what was particularly unique to media artists in the 1960s by comparison to early artists working with kinesis, was that movement and time were no longer used to explore the formal qualities of the artwork. Instead artists orchestrated movement to heighten the awareness of the relationships that are constantly reconfigured between viewers, each other, and the object during the experience of viewing. The de-emphasis of form was a necessary condition that focused on these new social and relational movements in time-based art.

What is significant to his argument in *Art, Action and Participation*, and his subsequent publications, is that Popper argued that artists who increasingly implicated the spectator/s in the aesthetic process and highlighted their movement affected the role and function of the kinetic object. For Popper demateriailty 'transfers the accent to the spectator...weakens the separate status of the object or "chef-d'-oeuvre", which is viewed no longer as an autonomous unit but simply as a stimulus or incitement to a particular type of activity or perception' (1975, p. 13). This has a significant consequence for Popper's consideration of the role and function of kinetic art because it is through participation and the dematerialised object that Popper suggested that the study of ontological movement ceased in the 1970s.

Therefore, while Popper considered that kinetic art was an important artistic practice in the 1960s because it had the potential to give rise to new art forms, he suggested that this was accomplished via the abandonment of creating objects of art that move. This is explicitly indicated when Popper stated that kinetic art was one of the earlier practices that orchestrated 'the public to participate effectively in transforming the existing environment' (p. 8). While also 'the disappearance of the object, the new role of the artist and the participation of the spectator, have a bearing on the aesthetic situation in general as well on kinetic art and the tendencies deriving from it' (1975, p. 232). For Popper, this 'disappearance of the object' was encouraged by the new forms of participation that were emerging, which rendered kinetic sculpture invisible, and decreasingly relevant to the explorations of new technological, media, and conceptual art.

For Popper, artists in the 1970s through to the present day who celebrated the materiality of industrial machinery in motion in an emerging digital age were no longer considered to be critiquing the technological society at the time, but instead tinkering with antiquated media. This is because, according to Popper, artists concerned with kineticism deferred to the popularity of conceptual art. As a consequence of this, Popper has suggested that conceptual and media artists after the 1960s were embarking on a 'post-kinetic' and 'neo-technological' state. (1987, p. 302). Even though in *Art of the Electronic Age* and *From Technological to Virtual Art* Popper explored the intersections between science, art, and new media, rather than old technological practices, in doing so he gives little room for discussing the developments of kineticism that have continued through to contemporary practices.

5.2 Technological Determinism

Popper's understanding and application of demateriality is distinctly different from Lippard's use of the term. It is important to note that while Lippard described demateriality as a separation of form and concept, and a de-emphasis on the material aspects of a work of art, materiality is still integral for understanding and discussing the conceptual elements of an artwork (vii). For Lippard, dematerialised conceptual art has inscribed in it a new relationship between form and concept, but this does not necessarily deem materiality as an unnecessary component of art.

By comparison, for Popper demateriality is a process that encouraged art to be temporal, relational, ephemeral, and lacking in form (1975, p. 278). His sense of demateriality is not only applied to identify a disjuncture between media and medium in postmodern and contemporary media art practices, but it is used to inform an historical intersection between conceptual and digital art. Consequently Popper builds a distinction between kinetic and 'technological artists' that focus on the effects of ontological movement to convey their ideas differently from artists using digital art and electronic media, which according to Popper, has been used to create immaterial, ephemeral and interactive art. This is how Popper has come to argue that artists have utilised demateriality to the point of entire material invisibility in *Art, Action, and Participation* (pp. 7–32).

Similarly to Susan Sontag's analysis of film, theatre, and participatory art in 'Film and Theatre' (pp. 24–37), Popper understood that art that unfolds in real time confronts kinetic art with a dual polemic: while movement in real time is a key defining element of kinetic sculpture, its use can potentially conflate kineticism with other practices that use movement , duration, and technology. Popper uses demateriality as a catalyst to blend participatory, conceptual, and technological art practices, and in doing so justifies his claim that kinetic art became increasingly indistinguishable from other practices (1975, pp. 7–11). Popper's study draws from a range of practices to develop and inform his approach to movement in art, such as automata, lumia, as well as robotic, cybernetic, and computer art, and provides an open definition of kinetic sculpture. In *Art, Action, and Participation* Popper conflates the distinction between kinetic and conceptual art and positions kinesis as the nexus between technological, scientific, conceptual, and cybernetic art:

> Kinetic art seems to have assumed the role of symbolically representing scientific and technical progress. It has shown the way towards the acceptance of electronic and cybernetic discoveries and their incorporation in the work of art (p. 7).

Participation in kinetic art soon became combined with participatory art, happenings, and public art (1975, pp. 26–32). In Popper's terms, '[t]hanks to new forms of participation [and] the dematerialization of the object through new technological procedures…the work as it was known traditionally is tending to disappear' (p. 278). It is from this state of disappearance that Popper suggests that kinetic artists turned towards more temporal and relational art like participatory events, public interventions, happenings, polysensorial environments, and theatre. This is precisely why Popper's appropriation of demateriality is used for both the defence and demise of kinetic artworks. While on the one hand demateriality points to the cross over between technological and conceptual art, particularly in the early stages of electronic art, it is on the other hand also used by Popper to suggest how and why artists ceased to practise a study of movement with movement. This is because for Popper a temporal approach to art helped to facilitate a diminishing regard for the kinetic object (1975, pp. 7–11).

While a post-kinetic approach was predominantly argued in *Art, Action, Participation* in 1968, Popper reifies this perspective thirty-one years later in *From Technological to Virtual Art*. In the latter text kinetic sculpture is framed as an important mechanical precursor to what he describes as ephemeral media such as holographic and multi sensorial digital installation, computer, virtual, and net art. Compared to mechanical media, virtual art is considered as, 'a new and refined version of technological art' (2007, p. xiii). Here virtual art is considered to be a 'new departure' from previous media art practices, and is characterised by 'innovation in visual and multisensorial perception, interactivity, and the development of aesthetic communication techniques' (p. 396). For Popper the components of digital media are more complex than the technology of prior decades, and they also consist of improved versions of earlier technologies. While this assumption aims to enrich the historical influences of contemporary digital arts, it does so by implying that contemporary mechanical media art is the antiquated modern counterpart to contemporary virtual art.

Popper's trajectory from modern kinetic art as a precursor to digital media positions virtual art as the immaterial heir and, in doing so, creates a distinctive digital determinism. Kinetic art is also discussed as having only rudimentary elements of immateriality, virtuality, and emphemerality (2007, pp. 1–8). This is because Popper applies the notion of demateriality to cast virtual media as the immaterial contemporary equivalent to modern kinesis art. Popper clearly explains that his method depends on media specificity when he states, '[t]echnological art was made up of several technically determined areas' (p. 7), that mostly described a modern mechanical, electronic art form.

In turn, the term 'virtual art' was drawn from a similarly technologically determinist viewpoint. As Popper considered that virtual art is the 'elements of all art made with the technical media developed in the late 1980s' (p. 7), this also excludes experiments with virtuality and digital art prior to the 1980s from being described as virtual art. This is affirmed when Popper continues that virtual art 'comprises not only of enduring digital-based work, multimedia off-line and online productions, and interactive digital installations but also what can be identified as a techno-aesthetic within these categories' (2007, p. 396). This perspective does not account for the continued desire to create and consume actual movement in artistic experiences, and instead constructs a media art history that nurtures a direct evolution from modern technological art towards digital and virtual practices.

This demateriality contributes to a binary association between kinetic mechanical sculpture as material, and digital and temporal art as immaterial. By describing virtual art as the 'refined version of technological art' (2007, p. xiii), Popper positions digital and virtual art as the more sophisticated and inevitable successor of kineticism. While of course there is ample cross over between late modern kinetic and digital art, Popper's approach alludes to an inevitable end to the desire to create and consume actual movement in sculpture and installation. If this were to be the case there would not be a continuation of kineticism in contemporary sculpture and installation, nor would there be an inclination to exhibit and connect with the modern kinetic sculpture.

This connection between demateriality and digital media has subsequently been echoed by a variety of art historians and critics. Edward Shanken has recently highlighted the use of technology in conceptual art as a direct tool for dematerialising the art object and reconsidering it as a process of information ('Art in the information Age: Technology and Conceptual Art' pp. 433–438). Similarly Julian Stallabrass has argued that information is 'offered in dematerialized form on the Web' (p. 61). Both Shanken and Stallabrass (2003) connect the notion of digitality with demateriality while analysing the intersections between conceptual and digital art. Additionally, Vladimir *Bonačić* (1989) has argued that cybernetic art is a specific medium that is used to dematerialise information (pp. 109–111). These considerations blur the distinction between digital processes, conceptual art, and demateriality only to consider kineticism as a part of a prehistory of contemporary digital media art and also converge demateriality with digitality in art.

5.3 Breaking Away from Binary Associations

However, this binary distinction is not necessarily always a consequence of dematerality in media art. Several critics and historians have considered the materiality of digital media in order to rupture the deterministic understanding of digital media as an immaterial and ephemeral tool. For instance, Jack Burnham became interested in the notion of a concept having 'material energy' when he curated the exhibition *Software* (1970). In an attempt to consider new metaphors for art in a new technological age Burnham drew together artists in the exhibition that highlighted informal as a material process (1980, pp. 200–215).

Burnham intended for *Software* to express post-formalist art as a culmination of conceptual, performance and interactive art that commonly separates medium from message in a state of demateriality. In the introduction to the exhibition Burnham explained that the curatorial choices of *Software* were intended to shape an idea of software, rather than hardware as a material process that is not necessarily abstract or immaterial (1970, pp. 10–14). He selected works that 'deal with underlying structures of communication or energy exchange instead of abstract appearances', some of which were programmed by computer systems, or like Haacke and Vito Acconci's performative works, by environmental and interactive situations. Burnham's idea of software was used to emphasise the presence of communication, concept, and process in conceptual art.

Similarly, albeit much later, Florian Cramer's (2004) *Words Made Flesh* offered an alternative critique to the relationship between medium and message in relation to information theory. Cramer considers that software 'is both material and practice' (p. 10), rather than supporting a binary association of software as immaterial and hardware as material both are comprised of material and immaterial assemblages. For Cramer, '[i]f the duality of software and hardware needs to be suspended, it follows that the notion of software as immaterial versus hardware as material must be suspended, too. The difference between materiality and immateriality exists within software itself' (pp. 10–14). Cramer determines that by neglecting to locate the material and immaterial properties that lie simultaneously within digital and analogue media, a rigorous scholarship of media art is prevented.

In a similar vein Mark Hansen (2004) has argued against an ephemeral image of digital information, and instead understands the body as a site for enframing information from both digital and analogue media (p. 11). Hansen rearranges the communicative structures in art through an embodied register and focuses on the specific relationship between the body and digital media is that 'digitization requires us to reconceive the correlation between the user's body and the image in an even more profound manner' (p. 10). In Hansen's view, the body gives form to information in both digital and analogue media equally, but in specific ways in each.

Shanken has also produced a useful categorisation of media that collapses binary associations between analogue and digital art. Rather than building a chronological account of technological experimentation in art during the twentieth century, early avant-garde artists are considered as players within Shanken's image of the contemporary art sphere. In his survey of early European avant-garde artists Naum Gabo and Laszlo Moholy-Nagy are considered in relation to contemporary public interactive trans-media works by Lozano-Hemmer, Horn, and Eliasson (pp. 55–77). Shanken considers avant-garde and contemporary artists to be a part of the contemporary interpretation of motion, duration, and illumination in art. This approach to media art history emphasises the continuation of motion, duration, and illumination in digital art and installation as an ongoing modern-contemporary conception, rather than as a set of practices that have only informed 'new' media techniques and effects.

There is a key and yet subtle difference in Shanken's approach to kinetic sculpture from Popper. The unfolding nature of movement and time under Shanken's view forms an ongoing history that runs parallel to the emergence and practice of digital art. For instance, Rafael Lozano-Hemmer's *Vectorial Elevation* (1999–2004), in Mexico City, included eighteen robotic searchlights installed in the inner city that were visible from a sixteen kilometre radius controlled by internet users from 89 countries. Shanken interprets the work as a piece of 'relational architecture' that disrupts the distinctions between the material as physical and definable and the digital as instant and virtual (p. 75). Light is used not only to signify communication, but it also collapses the zones of communication between the geographical space of Mexico and the digital interactions with it from all over the world. Lozano-Hemmer strives to heighten the awareness of time as an unfolding duration through both analogue and digital systems, because the instantaneous actions by users on the internet have actual, material, and durational effects.

Brian Massumi also considers Lozano-Hemmer's architectural light installations in a slightly difference way to Shanken, and yet also complicates assumptions that surround digital and analogue media. For Massumi, when decoding and recoding digital messages through light projection, as Lozano-Hemmer does in many of his works, the possibilities of merging and rationalising the invisible elements of analogue participatory art, communication, relation, and community are highlighted (Massumi 1998). This approach to digital media as a catalyst for emergence inverts Popper's argument by considering the ways digital media can visualise specific actions of communication.

Massumi's approach to emergence is informed by Gilles Deleuze's understanding of the virtual. Deleuze draws from Bergson's approach to the virtual as something that is not opposed to the material, but what is actual in the present. Consider for a moment a work like Hans Haacke's *Blue Sail* that was addressed in the previous chapter of this book. The sensorial and conceptual percepts that the viewer draws from the work are actualised in the process of perception. In this moment what the viewer does not perceive remains in a virtual state yet to be actualised in the moment of perception. Both the form and concept of the work are simultaneously loaded with virtual and actual properties in a process of actualisation that is in a continual state of becoming.

Even though Deleuze arranges the virtual and actual as being opposed to one another, they are also in a mutually exclusive dynamic. Massumi, through Deleuze, has emphasised the process of actualisation from the virtual as a continual mode of becoming where 'the virtual is the mode of reality implicated in the emergence of new potentials. In other words, its reality is the reality of change: the event' (Deleuze 1991, pp. 62–63). This raises the polemic within Popper's connection between virtuality, demateriality, and digital art because, through Deleuze, there is no 'increasingly immaterial' or dematerial. Rather, the virtual is something that enables actuality: '[t]he elements, varieties of relations, and singular points coexist in the work or the object, in the virtual part of the work or object, without it being possible to designate a point of view privileged over others, a centre which would unify the other centers' (pp. 62–63), the relationship between virtual and actual come together in a process of becoming, or emergence in the actual realm.

By comparison, Popper presents an understanding of materiality with a physicality that is concrete, visible, and determined. This highlights a paradox within Popper's logic because he overlooks kinetic artists who concentrated on the unseen relational and temporal processes that they articulated with electro-mechanical motion in the 1960s. While I go into more detail to critique Popper's use of demateriality later in this chapter through Anish Kapoor and Anthony McCall, this is also relevant to artists in *Directions of Kinetic Sculpture* such as Lye, Haacke, Bury, Robert Breer, and Tinguely, who were as much interested in the materiality of their works, as their orchestration of temporal and invisible elements of their work. The symposium for *Directions in Kinetic Art* consistently returned to the discussion of kinesis in art as having the potential to build a modality for the perceptual edge between visibility and invisibility, concern for the simultaneous material and immaterial elements, and the effects of movement.

Along with producing one of the most detailed historical analyses of kinetic sculpture to date in *Origins and Developments of Kinetic Art*, Popper's use of demateriality in *Art, Action, and Participation* holds considerable weight for the key understandings of kinesis in art history. However, as argued above, his argument that technological art has transitioned from mechanical and kinetic to 'more refined' digital media, such as virtual art, positions kinetic art as a precursor to other media practices, and which ceases as an art form when dematerial post-object aesthetics emerge. Popper only considers 'kinetic' according to movement, that is mechanical and material, which influences but distinctly falls short of digital art, even when movement, time, and process continue to be primary concerns for modern-contemporary artists. Popper's consideration of demateriality draws a new specificity between analogue and digital media. It is a problematic distinction that associates the digital with the immaterial and virtuality, while associating mechanical media with materiality and formalism. Not only does this conflict with Lippard and Burnham's use of demateriality but, also with other contemporary approaches to analogue, digital, and conceptual media.

5.4 Anthony McCall's *Line Describing a Cone*

Anthony McCall's *Line Describing a Cone* was originally exhibited at the Whitney Biennale in 1973, and has since consistently travelled to national and international exhibitions that have shown the work in relation to a variety of contemporary media discourses. The work was recently brought to the Australian Centre for Moving Image (ACMI) in Melbourne, in 2006 was at the *Eyes, Lies and Illusion* exhibition, and has also been exhibited in Sydney in 2005 through the Sydney Moving Image Coalition. Because of this the artwork has been interpreted from multiple disciplines including as a piece derived from expanded cinema (White 93–108), an avant-garde kinetic post-object oriented work, as well as a contemporary installation (Ballard 2008, p. 179).

McCall responded to questions about the formal aspects of the work and said that it 'sits deliberately on a threshold, between being considered a work of movement and being considered a static condition. Formalist art criticism has continued to maintain a stern, emphatic distinction between these two states, a division I consider absurd' (McCall 2003, p. 56). McCall explicitly created the work to evade formal categorisations of art and he does this by balancing liminal differences between the representation and presentation of movement with both digital and analogue media. Hence, the work can be discussed from multiple disciplinary modes, and which my interest in *Line Describing a Cone* lies in the use of movement in real time with the viewers.

Line Describing a Cone is a durational piece that is primarily constructed with film, projection, and movement of light. It begins in a dark space laced with fog that has been fitted with a screen at one end of the room and a projector facing the screen on the opposing wall. In the beginning a small white dot appears on the screen and for over the course of thirty minutes slowly forms into an outline of a circle. During this time a beam of light is sent from the projector to meet the formation on the screen. Over time the line transforms into the outline of a circle, which the projected beam follows, and an outline of a horizontal cone between the screen and the projector is produced. When left alone over the course of thirty minutes both projector and screen come together and enact this performance. Notably McCall created this work as an interactive installation and invites viewers to walk around and through the light projection.

During the work there is opportunity for the viewers to interact with the cone. They can choose to intersect with it by using their bodies to interrupt the light and create new formations, or stand aside to watch the cone grow into its complete formation. McCall uses the light beam to create relational movements between viewers, each other, and the light to recreate the space that they are in. In his words:

> For this film, every viewing position presents a different aspect. The viewer therefore has a participatory role in apprehending the event: he or she can, indeed needs to move around relative to the slowly emerging light form (1977, p. 53).

This has more recently been reiterated by McCall when he reflected, "[w]ithin the dark room, the individual audience members have to negotiate the space in relation to one another so that they can all see the light form" (2003, p. 44). It is important to note that while this work has not been often described as a kinetic work, its durational in nature and the movement of projected light moves in time and space with the viewer. Over time viewers gain a heightened sense of awareness of their own actions and the consequences of their movements: whether they are interrupting the light, potentially obscuring the view of other participants, and/or observing the formation of the cone. McCall orchestrates a duration that relies on the actual movement of communication and analogue media, and the two work together to negotiate the growing form of the light beam.

McCall's orchestration of movement is unique because the experience of *Line Describing a Cone* produces some friction with Bergson's approach to duration. This is because Bergson argues that film and photography reduce duration to a sequence of equidistant images that privilege a static image rather than an experience of movement (1913, pp. 321–323). The range of media that McCall uses is not used to highlight the different rhythms of motion between them. Rather, they coalesce with one another to provide a heterogeneous experience of duration: The projector's light beam moves in a continuous and unfolding motion, which meets the discontinuous illusion of movement created by the animated line on the screen, which is also altered by the interruptions to the beam made by viewers. McCall therefore draws together movement, projection, film, and installation to form movement that is simultaneously continuous and discontinuous, discrete and unfolding, and presented and represented.

When experiencing *Line Describing a Cone* participation and communication between viewers and the movement of the light beam counteract the representation of duration made by the animated line on the screen. This experience highlights both the continuous and discrete elements of the work at play together and creates an assemblage of mechanical and relational movement. It is nearly always an inevitable desire for viewers to come into the space and touch the light and break its path while the animation of the line on the screen continues to progress. The continuity between the projected light and the screen is a collaboration of two motions (light and screen) that are pulled apart and rejoined through the course of interaction throughout its duration.

One of the affects of *Line Describing a Cone* is that through each representation of movement, even if it simplifies duration, it is not mistaken for duration because the viewers touch, walk around, and interact with its progression in real time. McCall is adamant that the installation does not create a reduction of time but, conversely, heightens the awareness of time. This is evident when he says that '[t]his film exists only in the present: the moment of projection. It refers to nothing beyond this real time. It contains no illusion. It is a primary experience, not secondary: i.e., the space is real, not referential; the time is real, not referential' (2003, p. 42). This is one of the reasons why McCall created a work that, 'sits deliberately on a threshold, between being considered a work of movement and being considered a static condition. Formalist art criticism has continued to maintain a stern, emphatic distinction between these two states, a division [he] consider[s] absurd' (1977, p. 54). These formalist divisions of medium specificity distinguish the difference between dynamic and static works because they prevent new ways for articulating the experience of motion and time. McCall draws together multiple media to show that their conflicting presentations and representations of movement can build together an experience of duration that is contrary and multiple in order to heighten the awareness of the perception of movement in time.

Considering that Bergson's theses of movement in his time were largely based on reflections of how movement had so far been visualised in his lifetime, McCall demonstrates that technological expansion and assemblage of multiple media have the potential to produce new perceptions of movement and time. McCall's arrangement of digital and analogue media comes together to present movement and time in new ways that are different in kind. This touches on one of the key functions of kinetic art: to produce new perceptions and affects through the manipulation of time, space and movement.

Line Describing a Cone has been described as an ephemeral work, like many others involving the movement of light; it is a piece that is centered on the movement of time, and it lacks tactility and physical form, with exception of the screen and projector. Paradoxically, however, the experience of the light in motion has another effect. The light is not used to illuminate an area but is to cut through the darkened space and create a spatial dimension. Because of this, McCall explains, 'the more people who are present, the more "solid" the form becomes' (2003, p. 45). What gives the light as solid quality is the heightened awareness, attention, and interaction with the space that is generated over time. Interaction with the beam is nearly always an inevitable haptic desire, as bodies and limbs thrust through the wall of light drawing new formations with their shadows. In a crowded exhibition the risk of people interrupting the pathway of the projected light with their bodies becomes increasingly persistent. As the cone progresses, the preservation of the cone becomes a key objective for viewers.

Susan Ballard has suggested that one of the reasons why McCall's installation gains presence and becomes material and solid over time, is not only because of the relationships between the viewers, each other, and the work, but because McCall produces a form of materiality that is *emergent* (Ballard 2008, pp. 170–184). Rather than considering light as a weightless and immaterial tool, Ballard argues that McCall 'suggests an emergent materiality where fixed spatial objects generate unfixed spaces of questionable dimensions' (p. 46). For Ballard McCall therefore uses light to produce perceptions of dimensionality that grow in duration.

Ballard's observation, like McCall's description of the cone, interprets the beam as something that becomes increasingly solid over time and through movement. However, Ballard is more concerned with the generation of space, rather than the attention and interaction with the light itself. The emergent materiality that Ballard sees is not a condition unique to McCall but, uses *Line Describing a Cone* as an example to develop an alternative understanding of analogue and digital media that diverts from binary associations that solely consider digital media as immaterial and analogue media as material.

For Ballard, emergent materiality is one of the unique affects of both digital and analogue media in installation (pp. 53–62). Her interpretation of emergence is influenced by Deleuze's process of becoming that lies between the virtual and actual, and she uses it to foreground the qualities of becoming and intensity in the moment of actualisation. Ballard shifts traditional definitions of material and materiality away from what physically constitutes the artwork, towards matter in-formation.

Ballard's approach resonates with Florian Cramer's perspective of information as a process (pp. 121–125), as well as Matt Kirschenbaum's understanding that digital information is a 'function of the material and historical dimensions that obtain for *all* artefacts'. For Ballard, digital materiality points to the affect of the medium—digital materiality is what digital media *does*, rather than what it is: '[t]he digital *does* this by mutating its very surfaces and interlacing these with a viewer, so much so that the digital image is no longer tied to an external reality, but to the processes of viewing' (p. 67). This is a distinct shift away from Popper's chronological and media specific approach to defining digital and technological art because Ballard concentrates on the material and immaterial affects of media in art.

By collapsing modes of media specificity Ballard draws from Peter Lunenfeld's (2000) *The Digital Dialectic* to consider that all information, whether it be digital or mechanical, can be stored, distributed, accessed, and altered (although to varying degree). This variability suggests that information in both analogue and digital media depends on an arrangement of material and immaterial processes folding together, and directly conflicts with Popper's digital determinism. For Lunenfeld, digital materiality 'encompasses processes, which intersect with analogue notions of matter information. These arrangements are not composites or sums but assemblages' (p. 66). The 'informational field' therefore looks to material and digital information as a form of processes that come together, both in material and immaterial forms.

Ballard's approach to digital and analogue materials is useful because of the way she veers away from technologically determinist media specificity by disregarding the binary association between the digital as immaterial and the analogue as material, as drawn by Popper. Ballard sees that the virtual is not equal to the digital but refers to the realm of potential that could emerge in the present.

From Ballard's position all media have a variety of qualities within them that are binary and open. Ballard's emergent materiality creates an opening for considering the emerging processes that occur through digital and analogue assemblages that affect viewers in artistic experiences. In this respect Ballard is useful for approaching lumia, screen based, and participatory art, as tools for invoking and/or creating a spatial emergence that is both material and immaterial, rather than increasingly invisible as Popper suggests.

Although Ballard's argument is not restricted to kinetic installation I would like to suggest that temporality is key to the emergent materiality that she gauges in *Line Describing a Cone*. According to Ballard, '[t]his film installation is periodic although the experience of it is not necessarily so. The emergence of the film is continuous, although the experience of the installation is distributed, interrupted and distorted. It flickers between analogue and digital' (p. 182). The tensions of this flickering are enabled because the installation is orchestrated by multiple and, at times, conflicting temporal rhythms of various mechanical media and viewer participation. The installation slips between formalist categorisation, and is able to flicker between discrete (digital) and open (analogue) effects partly because it is a time-based work.

If McCall constructed the cone as an entirely static form, its affects would be eschewed considerably. By orchestrating time with movement McCall is able to produce formal (material) and conceptual (immaterial) processes. Therefore, through movement, the concept is not dematerialised, but reaffirmed through the durational unfolding of its form. In doing so, in light of Ballard, material and immaterial elements of the work can be emergent rather than immaterial and ephemeral.

In this sense *Line Describing a Cone* can be used as an example to destabilise digital determinist view such as those expressed by Popper. Firstly, Popper's approach to digital media assumes that it is predominantly considered as an ephemeral, invisible, and immaterial medium (1975, p. 278). However, McCall reminds us that a the movement of light in installation can hold material and immaterial processes simultaneously through an orchestration of discrete, continuous presentations and representations of movement. Secondly, the form and concept of the installation is assembled to develop a materiality that is emergent over time. The formation of the cone not only occurs over time but, it is through time that awareness and attention of its materiality is heightened. Rather than considering that kinetic sculpture is becoming increasingly invisible, McCall uses light in movement to generate visible dimensionality and locate space.

Rather than considering progression from mechanical to digital media, McCall's work is useful for exploring the material and immaterial effects of technological media. Rather than an entirely formless system of movement and communication between the artwork and the viewers, McCall orchestrates form and participation together in order to highlight that movement in art involves the assemblage of material and immaterial qualities through the orchestration of actual movement.

5.5 Movement as Emergent and Incipient

Another technique for artists to elaborate on the emergent nature of movement in art is by arranging movement for viewers to feel for its rhythms during its incipience. Anish Kapoor's *Shooting into the Corner* (2009) is a simple mechanical installation that consists of few elements: a canon that is loaded with pre-fabricated canisters filled with scarlet wax. At twenty-minute intervals an assistant fires the ammunition that shoots through a doorway and against the opposing white wall of the gallery space.

During the exhibition viewers of the installation are invited to stand in a partitioned area and view the performance of the canon firing. At a scheduled time an assistant is directed to enter the room and reload the canon with an eleven-kilogram container of new wax and fire it by pulling a mechanical lever. Unlike many of his more minimalist works, Kapoor's *Shooting into the Corner* is an experience abundant with excess. The canon itself fires with a large bass frequency that echoes throughout the gallery halls, and over time the waste of the fired wax canisters accumulates across the walls and floor of the space. Over the duration of the exhibition a mountain of wax builds under the firing zone, while its surrounding areas become splattered with red remnants.

Shooting into the Corner is as much an orchestration of matter as it is of duration. Various videos that document the experience of the work in multiple exhibitions show audiences arrested by silence during the intervals between loading and firing the canon and often jeering, celebrating, and applauding after it fires. The moment of time that each attendant takes between loading and firing the canon varied, some waited longer than others and let the uneasy expectation of their audience build, while others pause momentarily before pulling the lever (Dodgson 2010, p. 763). In this moment the perception of time passing is heightened as expectation builds because expectation is manipulated to give into uncertainty.

Indeed, it is when we think of a moment that we create an isolation of time, a discontinuity of rhythm. This discontinuity is not a cut in time but, perception is heightened in a matter of degrees and intensity, and the awareness of time is isolated to a similar degree. The moment when audience members are held in suspense before the canon fires in an instant that passes too quickly—it is the moment prior to it that excruciatingly extends time, draws it out, and heightens our perception of its passing.

These audience members react to a building of suspense in a similar way to the viewers of Jean Tinguely's *Homage to New York* (1960), which I addressed in detail in Chap. 2. When reflecting on Tinguely's performance in the garden of the Museum of Modern Art, Billy Klüver describes feelings of rising intensity and suspense as the work destroyed itself in ways that were unforeseen by the artist (Klüver 1987, pp. 74–77). The suspense that was felt during Klüver's experience, however, was due to the accumulation of uncertainty. A similar reaction is orchestrated by Kapoor's kinetic installation, however the suspense is accumulated, predicted, and relieved through unpredictable repetition.

Suspense is often orchestrated when the progress of information is withheld and/or uncertain. Even the threat of uncertainty in film and literature can be contributors to suspense (Yanal 1996, pp. 146–158). Klüver's feelings of suspense were symptomatic of the uncertainty of how the auto-destructive performance would end. Yet, unlike Klüver, the audience members viewing *Shooting into the Corner* feel a rising intensity and suspense before the canon is fired, even though the installation is predictable and repetitious in nature. When they walk into the exhibition space for the first time they are already confronted with an accumulation of red wax along the walls and floor that serves as evidence for what they will witness. The canon also fires at set intervals, provided that the assistant fulfills his/her requirement (which they inevitably do). Despite the predictability of the canon's fire, audience members wait in silence and perceive with building tension and, as noted previously, release with cheers after the wax pellet is fired.

These reactions to the performance of movement are tied to the perception of an uncertainty around when the canon will be fired rather than how or if it will happen. This partly explains how Klüver's perception of duration felt as though it became elongated to a point of excruciating length, even over an actual moment in time. I am suggesting that not only is suspense created by narratological ambiguity, but also these feelings of suspense signify the intensity of a moment accumulating before it is actualised in the present, particularly in temporal works such as Kapoor's.

One way to approach this idea is to consider the perception of movement in time as something that builds. It is emergent, as Ballard suggests, but it may also emerge from a point of immanence before it comes into actualisation. Erin Manning's (2009) *Relationscapes: Art Movement, Philosophy* builds a framework for what she terms 'preacceleration': the primary phases of movement in its initial incipiency. Manning gauges the rhythms of perceiving movement in order to understand how its intensity builds before it is present. This preacceleration is a way of seeing the temporal present as a continually productive force that territorialises the here and now.

Rather than thinking about static sculpture as an ideal representation of movement, Manning draws from Deleuze's movement-image to consider sculpture as a form of mobility-becoming-image (p. 131), rather than as a static image of movement. Manning explains:

> Although the sculpture seems to give into the pose, the pose is in fact given to the sculptural movement. You create the pose by prehending it. You actualize it. But its actualization is only as real as the intensive movement that backgrounds it. Any actualization can produce a pose, but this pose will have been created by a mobile cut that is a becoming intensive of extensive movement. The pose is never the starting point (p. 131).

Manning, via Deleuze and Bergson, arranges of mobility-becoming-image, and refers to the movement that is implied between representations of movement in film, or the 'prehension' of the static pose in sculpture (p. 131). Mobility-becoming-image also articulates movement before it is actualised as a form of preacceleration. If actual form, like the sculptural pose, is not the starting point of movement, then in time-based kinetic works such as *Shooting into the Corner* actual movement of the installation is also not the starting point of mobility.

Manning's Deleuze-Bergson approach to movement engages with duration as an eternal and material process, rather than something that is immaterial. Duration is considered as a form of unfolding eternity, which when interpreted and visualised creates new rhythms. This approach to deconstructing motion in contemporary art lends itself to a reinterpretation of modern time by borrowing from Bergson's universal (yet multiplicitious) duration (pp. 37–49). This time is a continual unfolding and is unified in the actual present, and what remains to be actualised resides in the virtual domain. What is unique to Manning's argument is the way in which time and space are not approached as opposing elements of kinetic art, but rather work together in synthesis within the intricacies of the virtual intensifying and informing the preacceleration of the actual.

The very anxiety of predicted movement is the mechanism that prolongs Kapoor's moment before motion. While the action may be instantaneous, anticipation intensifies the intuition of preacceleration and perception of actual motion. This is how Kapoor manipulates both time and space: rather than considering them in opposition to one another, instead they are used in synthesis in order to create an entirely predictable moment that is also pregnant with anticipation. In relation to *Shooting a Canon into a Corner*, he states:

> I am interested in sculpture that manipulates the viewer into a specific relation with both space and time. Time, on two levels: one narratively and cinematically as a matter of the passage through the work, and the other as a literal elongation of the moment....Space is as complex. The space contained in an object must be bigger than the object that contains it. My aim is to separate the object from its object-hood (quoted in Reitmaier 1965, p. 92).

Kapoor's use of actual movement manipulates his audience to feel-with the intensity of motion before it occurs, which is what creates a sense of suspense. The audience feels for, predicts, and searches to catch the moment of actualisation before it comes. This elongation of the moment is also met with an immediate passing of its instant; the audience waits in silence, gathers their perception of duration, to be shocked with the short moment that it takes to fire the pellet and for it to hit the wall. The accumulation of anticipation stretches the moment, slows time to feel the precise moment of present that finally passes with a cathartic explosion of the canon.

Ballard and Manning provide a vocabulary for articulating the perceptions and reactions manipulated by Kapoor in *Shooting into the Corner*. Through Ballard and Manning movement is not only an emergent process, but that very emergence is created from a building state of intensity between the actual and virtual. Preacceleartion shows that this emergence comes from the building of intensity in the virtual becomes singular and actualised in the present. Kapoor makes his audience feel this exactly by building the perceptions of suspense, anxiety, and celebration that develop when his wax is fired in the gallery.

Through Ballard and Manning, it also becomes evident that movement cannot only be emergent but, is a process of emergent materiality from incipience. This stresses the simultaneous qualities of movement as something that is virtual and actual. Similarly, duration is both material and eternal. *Shooting into the Corner* is a time-based kinetic work that formally presents a similar understanding of movement. Kapoor's simple gesture of movement mediates time: The moment of action is 'elongated' to encompass not only the actual action of the canon firing, but the sense of it occurring before it does.

Viewers of the artwork are manipulated to see and feel the explosion and vibration of the capsule of wax firing out of the canon, but also sense the arrival of its actualisation, or what Manning would describe as developing a sense of preacceleration. The manipulation and elongation of this eventual moment that is entirely unpredictable and violent. Yet Kapoor's canon behaves normally for a canon: It repetitively shoots capsules of wax that eventually build against the walls and floors of the gallery. The repetition of its action, despite being unpredictable, is precisely the mechanism that draws out the perception of an emergent moment of materiality.

5.6 Materiality, Participation and Anticipation

Frank Popper's approach to kinetic art in relation to conceptual and digital art has formed an influential, and yet polemical construction of kinetic art history. More specifically, Popper's interpretation of the post-object tendencies in time-based art since the late 1960s has serviced an assumption that considers demateriality as a catalyst that enabled kinetic art to be absorbed by digital art practices. Popper's argument justifies this transition by associating kinetic sculpture with the weighty industrial materials of modern art, and conceptual and digital art as communicative, ephemeral, invisible, and relational. I have approached Popper's argument as a key polemic in contemporary art history because it foregrounds a media specificity that suggests that mechanical media is predominantly an historical precursor to digital media.

Additionally, in this chapter 1 have highlighted the works by McCall and Kapoor, and the concepts of emergence and preacceleration by Susan Ballard and Erin Manning to emphasise the polemics in Popper's perspective. I argue that these artists celebrate materiality through kinesis in art on a variety of scales, whilst engaging with temporal relational processes such as participation and anticipation. *Line Describing a Cone* has been highlighted as an installation that consists as a collaboration of time and movement rhythms that creates space as it unfolds. Lumia, projection, and film are used to find a perception of movement that is ephemeral and yet emergent as well as being continuous in time, that unfolds with an array of discrete discontinuities. My analysis has also drawn from Ballard and Manning, and to consider movement (regardless of media specificity) as an incipient and emergent process that produces both virtual and actual affects. Kapoor's *Shooting Into the Corner* uses kinetic movement because it is a medium that is able to manipulate the perception of duration. Movement is used for its incipient and excessive qualities in Kapoor's work. It builds in intensity and is excessive in its means, signifying a continued desire to use movement in contemporary art to manipulate perceptions of duration.

My analysis of Frank Popper's historical understanding of kinetic sculpture and the contemporary works by McCall and Kapoor also contribute to Smith's understanding of contemporary art and conceptualisations of contemporaneity in multiple ways. First, Popper's inclination to consider kineticism as a mechanical, machinic technological precursor to virtual art strengthens a notion that kineticism is in many ways a form of modern art that is distinctively different from the practices of postmodern and contemporary technological media. In this, Popper provides an opening for exhibitions such as *Revolutions—Forms that Turn* to be interpreted as returning to and reassessing modern art practices. Popper therefore contributes to a contemporary art historical perspective that facilitates a remodernist view of contemporaneity in contemporary art.

Secondly, even though I have used arguments by Manning and Ballard to destabilise Popper's digitally determinist understanding of the relationship between kinetic and virtual art, I have turned to the works by McCall and Kapoor emphasise the material and immaterial effects of motion, while also privileging a universal present that is also heterogeneous in nature. This also resonates with Smith's claim that contemporary artists are often fixated on the contrariness of how temporality is conceptualised today (Smith, 'To Be with Time is All We Ask' 309–315). This is evident in the way McCall orchestrates his work to oscillate between discrete and continuous movement and evades to also produce a unified present. Kapoor also draws together modernist tropes of temporality: rhythm, repetition and incipient time. My analysis of McCall and Kapoor also raises questions about the material, immaterial, actual, and virtual effects of kineticism in general.

The following chapter will address early experiments with kinetic sculpture as an exploration that oscillates between these aspects of perceiving movement through the works of early kinetic artists, namely László Moholy-Nagy and Naum Gabo. Like McCall and Kapoor, I argue that these early European avant-garde artists used movement to encourage the viewer to consider mechanical movement as a material and immaterial process.

References

Ballard S (2008) Out of order: explorations in digital materiality. Doctorate Thesis, College of Fine arts at the University of New South Wales

Benthall J (1972) Science in art and technology today. Praeger, New York

Bergson H (1913) Creative evolution, (trans: Mitchell A). Macmillan, London

Bonačić V (1989) A transcendental concept for cybernetic art in the 21st century. Leonardo 22 (1):109–111

Brett G (2000) Force fields phases of the kinetic. Hayward Gallery, London

Burnham J (1968) Beyond modern sculpture: the effects of science and technology on the sculpture of this century. G. Braziller, New York

Burnham J (1970) Notes on art and information processing', software, information technology: its new meaning for Art. Jewish Museum, New York

Burnham J (1980) Art and technology: the panacea that failed. In: Woodward K (ed) The myths of technology; technology and postindustrial culture. Routledge, London

Cramer F (2004) Words made flesh, code, culture, imagination. http://www.netzliteratur.net/cramer/wordsmadefleshpdf.pdf. Accessed 9 Apr 2015

Deleuze G (1991) Bergsonism, (trans. Tomlinson H. and Habberjam B). Zone Books, New York

Dodgson N (2010) Sculpture: engineering art. Nature 468, pp. 762–76. http://www.nature.com/nature/journal/v468/n7325/full/468762a.html. Accessed 29 Feb 2016

Fried M (1998) 'Art And Objecthood', Artforum, 1967; reprinted in art and objecthood. University of Chicago Press, Chicago

Hansen M (2004) New philosophy for new media. The MIT Press, Cambridge, Massachusetts

Klüver B (1987) The garden party. In: Hultén P (ed) A magic stronger than death. Thames and Hudson, Venice
Lippard L (1997) Six Years: the dematerialization of the art object from 1966–1972, 2nd edn. University of California Press, Berkeley
Lunenfeld P (2000) Introduction. screen grabs: the digital dialectic and new media theory. The digital dialectic: new essays on new media. The MIT Press, London
Manning E (2009) Relationscapes, movement, art, philosophy. The MIT Press, Cambridge, Massachusetts
Massumi B (1998) Sensing the virtual, building insensible, Hypersurface Architecture: Architectural Design 68, vol. 5, no. 6, pp. 16–24,
McCall A, Dasgupta G (1977) 'Interview: formalist cinema and politics'. Performing Arts Journal 1(3):51–61
McCall A (2003) Line describing a cone and other films. vol. 103, pp. 42–62
Popper F (1968) Origins and developments of kinetic art. New York Graphic Society, New York
Popper F (1975) Art, action and participation. Studio Vista, London
Popper F (1987) Editorial: technoscience art: the next step. Leonardo 20(2):301–303
Popper F (2007) From technological to virtual art. The MIT Press, Cambridge, Massachusetts
Reitmaier H (2002) 'Descent into Limbo', Tate international arts and culture vol. 1. http://www.tate.org.uk/context-comment/articles/descent-into-limbo-anish-kapoor. Accessed 06 Jun 2017
Rosenberg H (1965) 'The American Action Painters', Art News 1952; reprinted in The Tradition of the New. McGraw Hill, New York, pp 23–39
Shanken E, Ascott R (2003) Telematic embrace: visionary theories of art, technology and consciousness. University of California Press, Berkeley
Stallabrass J (2003) The aesthetics of net. art. Qui Parle 14(1):2003
Yanal RJ (1996) The Paradox of Suspense. Br J Aesthetic 36(2):146–158

Chapter 6
Three Key Influences of the Kinetic Aesthetic: Henri Bergson, László Moholy-Nagy and Naum Gabo

I suddenly saw the difference between concept and reality
(Sibyl Moholy—Nagy 'The Light Display Machine'
Kostelanetz 1971, p. 147)
Movement does indeed exist here; it is in the apparatus
(Bergson 1913, p. 322)

Abstract This chapter presents László Moholy-Nagy, Naum Gabo, and Henri Bergson as three key influences of the roles and functions of movement and time in kinetic art history today. It begins by highlighting Henri Bergson's concept of duration as the perception of the present as it unfolds, and the memory of the past affecting the experience of the present. While Bergson argued that any mechanical representation of movement produced a distortion of duration I highlight that Moholy-Nagy and Gabo used the movement of mechanical media to explore the relationship between movement and machinery to produce new perceptions time. I argue that the use of movement, acceleration, repetition, modulation, and the optical effect of 'virtual volumes' with light and machinery enabled Moholy-Nagy and Gabo to articulate the present as an accumulation of temporalities as they unfold which expand the range of temporal codes within duration.

The following chapter concentrates on two artists László Moholy-Nagy and Naum Gabo, and philosopher Henri Bergson and three key figures who have influenced the role and effects of movement, time, and technology in kinetic sculpture. More specifically in this chapter I address Moholy-Nagy and Gabo's approach to mechanical movement to complicate Bergson's attitude towards the relationship between experienced time and technological mediation of movement. Rather than distorting or simplifying the experience of duration, I highlight that Moholy-Nagy and Gabo use kinesis in their works as a balanced experimentation of actual movement and expressing the sense of time passing through the present. Both artists produce new spectacles of duration through the mechanical production and reproduction of movement. While these three figures are remembered as modern pioneers of movement in art and philosophy, it is important to note that the role of technological movement remains a key question in art today, or as Ernest Edmonds

© Springer Nature Singapore Pte Ltd. 2017

C. Chau, *Movement, Time, Technology, and Art*, Springer Series on Cultural Computing, DOI 10.1007/978-981-10-4705-3_6

has claimed "We are being swept forward in an accelerating cultural revolution of unprecedented scale" which had influenced the understandings of movement, time and technology in art (p. 92).

This chapter will begin by focusing on Henri Bergson's critique of modern Western representations of movement in relation to duration. I will highlight that rather than defining time through space and dissecting it into quantitative units, Bergson's duration, by contrast, is characterised by continual transformation. The lived experience of time constitutes a process of constant bifurcation from the present (actual) to the past (virtual) while the very awareness of this continual change itself participates as its own rhythm of duration.

From Bergson's perspective the mechanical representation of movement is problematic because, 'we give a mechanical explanation of a fact, and then substitute the explanation for the fact itself' (Bergson, *Time and Free Will*, 1910, p. 181). For instance, the reductive representation of movement in film is mistaken for a presentation of actual movement in duration. For Bergson mechanical movement depends on discrete, isolated, and repetitive actions that differ from the qualities of the subjective experience of time. Therefore Bergson's duration as a perception of temporality that focuses on movement as a process that is not subservient to images of time, but contributes to time as an unfolding present.

Secondly, this chapter will approach Moholy-Nagy's (1965), *Light Space Modulator*; *Vision in Motion,* and *The New Vision,* as key texts that indicate the artist's approach to the presentation and representation of movement in modernity. Moholy-Nagy's approach to movement differs from Bergson's because the artist uses the presentation and representation of mechanical motion to expand perceptions of time. Moholy-Nagy's kinesis is therefore a means for exploring the creative productive modes of technological media that defer from notions of mechanical motion as reductive, binary, and repetitious. Many of Moholy-Nagy's photographic experimentations focus on altering perceptions of the present to stretch across the frame, rather than presenting it as an instant. Images of time are produced as continuous durations that are preserved within the frame.

Similarly, the movement of light as a process that is continuous, but also modulating, refracting, and rejoining, all of which are perceived when the viewer moves with and around *Light Space Modulator*. One of the unique aspects of the work is the way Moholy-Nagy demonstrated the differences between representing movement in film and photographic media, to various settings in which the sculpture itself could be experienced. Because of this, I argue that mediation of motion does not simplify duration, but has the capacity to produce new perceptions of its rhythms.

Thirdly, Naum Gabo's *Standing Wave: Kinetic Construction* (1920) is presented as a direct expression of the key claims found within his *Realist Manifesto* (1920). Both the manifesto and sculpture indicate Gabo's exploration of the limits of human perceptions through kinetic dynamism in real time. Gabo's approach to kinesis in art is an orchestration of real movement in real time that is used to privilege the present as an unfolding and continuous event. Gabo resisted the avant-garde tendency to focus on art as a means for bringing attention to the future, and instead used movement to draw attention to the present temporality.

Gabo's 'virtual volume' is an important optical effect of movement in *Standing Wave*. This volume is an optical illusion that is produced when an object moves in rapid repetitive motions and creates a blur of present and past perceptions. Gabo's kinesis in sculpture is, then, an orchestration of the perception of actual and optical (or virtual) movements. The perceptual edge between material and optical volumes enables movement in the present and impressions of past motion to be consolidated while a viewer perceives the work. For Gabo, kinesis is not only used to explore the perceptions of the movement, but also to form a perception of temporality that is plural, folding, and as a constant transformation.

Bergson has been an influential philosophical figure for the artists and historians who orchestrate actual movement in their art. This is partially because many of these artists explore the use of actual motion, rather than its representation, in order to articulate the nature of how we perceive ourselves perceiving motion and time in artistic experiences. My analysis of Moholy-Nagy and Gabo's use of kinesis problematises Bergson's resistance to mechanical representations of movement in a number of areas through the perceptual effects of mediation, modulation, and the virtual volume.

6.1 Resensationalising Early Modern Kineticism

Moholy-Nagy's *Light Space Modulator* ((1922–30), also known as *Light Prop for Electric Stage* and *Lichtrequisit Einer Eelektrischen Bühne*) and publications have been crucial to art historical discussions in recent decades in connection with movement, time and technology in contemporary art. The kinetic sculpture has experienced a renaissance in recent years, since the artist's daughter, Hattula Moholy-Nagy, in 2000 authorised a reconstruction and donated it to Harvard Art Museum specifically for their Busch-Resinger Museum collection. The commission of the full-sized replica stipulated that after the replica was made in 2006 it must be lent to major exhibitions whenever possible, and that the Tate Gallery had the right to display the work one out of every fours years.

Even though the Bauhaus Archiv Berlin and Van Abbemuseum in Eindhoven already own earlier replicas of the *Light Space Modulator*, the agreement between Busch-Reisinger and Tate Museum has encouraged a renewed analysis of the work in relation to contemporary discussions in art and society. Subsequently the artwork has appeared in a number of exhibitions such as *Moholy-Nagy and Joseph Albers: From the Bauhaus to the New World* (2006), *Light Display Machines: Two Works by Moholy-Nagy* (2007), at the Busch-Reisinger Museum; *Moholy-Nagy in Motion* (2011), at The National Museum of Modern Art, Kyoto; as well as *Moholy-Nagy: An Education of the Senses* (2012), at the Loyola University Museum of Art; *Museum Modules: Play Van Abbe, Part 2: Time Machines* (2010), at the Vanabbemuseum, and an online exhibition titled *Extra Ordinary Every Day: The Bauhaus at the Busch-Reisinger* in 2005. The construction of the replica and circulation of a major international exhibition has allowed the work and the artist's work in general to 'get a new lease on life' (Nisbet 2012).

Since 2006 The *Light Space Modulator* has been regarded as ‘a classic work of Modern art’ (Nisbet 2012), ‘a seminal kinetic sculpture’ (Harvard Art Museum), a work of ‘transformative vision’ (*Lyola University Museum of Art*), a symbol of modernity (*Kunsthalle Erfurt*), and a pioneer of contemporary understandings of modulation, duration, and movement of light (Shanken, *Art and Electronic Media*, 2009, pp. 55–56), and Moholy-Nagy has been deemed a ‘mentor to modernism’ (Englebrecht). The positive acclaim for the work itself has flourished more so recently than before, when the artist was alive. In Moholy-Nagy’s time the work was criticised as an obscure and ambiguous work and a dated fetishisation of a constructivist machine aesthetic (Tsai 2011, p. 293).

In 2011 Naum Gabo’s *Kinetic Construction: Standing Wave* featured in an exhibition with Moholy-Nagy’s *Light Space Modulator* titled, *Immaterial: Brancusi, Gabo, Moholy-Nagy* (2004) at Kettel’s Yard, Cambridge. While Gabo’s *Kinetic Construction: Standing Wave* has not experienced a recent reception as popular as *Light Space Modulator* the artwork continues to be considered as a key modern art historical piece within contemporary art historical discussions.

The sculpture has been recently considered as a work that is central to the artist’s aesthetic (Hammer and Lodder 2000a, b, p. 69), and a pioneering example of using movement as an artistic medium (Popper, *From Technological to Virtual Art*, 2007, p. 12), even though it is Gabo’s only known kinetic sculpture. While the history of the work’s reconstruction and construction of replicas has not spurred as much contemporary discussion as Moholy-Nagy’s *Light Space Modulator*, a replica of *Kinetic Construction: Standing Wave* is a part of the permanent collection exhibition of the Berlinishe Gallery in Berlin *Art in Berlin 1880–1980*.

While art historians Terry Smith and Pamela Lee have considered the 1960s as an influential decade for the contemporary interpretations of time in art, exhibitions such as *Revolutions-Forms that Turn* frame avant-garde artists during the early decades of the twentieth century as key figures for art history in contemporary society. My earlier interpretation of the curatorial arrangement of *Revolutions-Forms that Turn* in Chap. 1 positions the contemporary contemplation of time and contemporaneity as being influenced by artists of the early European avant-garde. Chapter 1 of this book addressed the role and function of avant-garde kinetic sculpture in contemporary exhibitions as a means for remodernising a modern machine aesthetic in art theory, history, and criticism today. It is, therefore, also important to consider the use of movement and time in early avant-garde kinetic experiments as influences of contemporaneity in contemporary thought.

The technological expansion of the industrial revolution in Western society brought new discussions around how space and time were perceived in modern media such as film and photography (Solnit 2003, p. 11), and therefore came to characterise what Althusser and Balibar would describe as a form of peculiar time in modernity (99). For many, these elements underwent a process of annihilation. Time, rather than space, was a key consideration for modern art and philosophy.

Rather than equating time to space, or validating the distortions of temporality perpetuated by Western culture, Bergson discussed time as an event that cannot be represented and is constantly bifurcating between present and past between its actual and virtual form. Similarly, Moholy-Nagy and Gabo created their kinetic sculptures in order to seek new ways to express time that did not distort or simplify its experience through representation, but explored the limits of perception through actual and illusory forms. Actual movement in real time was for Moholy-Nagy and Gabo an effective means to explore the limits of perception of movement. Movement was not only an expression of the present temporality but, also a tool for considering the act of perceiving time as an intuitive and malleable process.

6.2 Bergson: Modern Western Movement, and Duration

In *Time and Free Will, Matter and Memory,* and *Creative Evolution*, Henri Bergson unpacks three dominant ways in which Western society rationalised and represented movement and time. Throughout these texts the representation of movement in film, photography, and—to a degree—painting are critiqued for distorting the experience of time. For Bergson, these representations separate time into equidistant static images and spatialise its form. For instance, Eadweard Muybridge's photographic motion studies of a horse galloping along a racecourse is problematic because it reduces the animal's movement into a series of instances, to which movement occurs *between* points A and B as a subordinate process to the image represented. As Bergson notes, 'all division of matter into independent bodies with absolutely determined outlines is an artificial division' (*Matter and Memory,* p. 259). For Bergson, this is because movement is a qualitative transformation that is distorted when divided and quantified.

Bergson explains that representations in modern media according to a series of instances (such as film, photography, animation, and zoetrope), or ideal images (representations of growth or movement presented in stages of change over time) are problematic. Bergson's reasoning for this is that not only are these representations a distortion of duration, but also, and more importantly, 'we give a mechanical explanation of a fact, and then substitute the explanation for the fact itself' (*Time and Free Will,* p. 181). Modern media is therefore characterised by Bergson as conceptualising and expressing time in a spatialised and quantifiable manner. Within these representations time exists in an interval between images of movement, and is represented in a repetitious, linear, and mechanised fashion.

In this regard representations of movement fail to demonstrate that time is subjectively experienced as constantly changing. In Bergson's words, 'what is real is the continual change of form: form is only a snapshot view of a transition' (*The Creative Mind,* p. 15), and to create a snapshot of movement is to represent it exactly as it is not: static and unchanging. Bergson describes this static image of movement as an interruption of actual movement because 'a passage is a movement and a halt is an immobility. The halt interrupts the movement itself' hence exposing the contradiction of representation (*Matter and Memory*, p. 247). For Bergson, time becomes spatialised through the mechanisms of static representation that signify transformation. This is particularly problematic because to render time dependent on space also reduces it to something that is divisible, measurable, and homogenous. Whereas Bergson considers the representation of movement inherently incapable of creating a true experience of duration. Through film, representations of movement can be problematic as they can be mistaken for actual movement rather than be recognised as illusionary. The increasing, everyday consumption of these illusions of movement make the risk even more prevalent.

For Bergson, duration is the time that we experience subjectively. It not only entails the perception of the present as it unfolds, but it is also the memory of the past affecting the experience of the present. Therefore, the present preserves images of the past while gathering the potential future, both of which coexist virtually with the present. In this sense, duration is both a continuous unity while also heterogeneous in nature (Ross 2012, p. 23). In *Time and Free Will,* for example, duration is described as: 'On the one hand, a *multiplicity* of successive states of consciousness and, on the other hand, a *unity* which binds them together' (Deleuze, *Bergsonism,* 1991, p. 45). This unity consists of the actual events that form in the present. As Bergson continues, '[d]uration will be the *synthesis* of the unity and multiplicity' (Deleuze *Bergsonism,* 1991, p. 45). Therefore, duration inherently consists of change, transition, and becoming, and its rhythm in the present—as well as the past preserved within it—is inherently unpredictable, and continuously in a state of becoming.

To help explain the concept of duration, Bergson describes the act of watching sugar dissolve into a glass of water in *Creative Evolution*. The process as a whole involves the weaving of interactions between the glass, water, and sugar, as well as Bergson's own subjective perception; his senses, mind, patience, and boredom. In his words, 'the time I have to wait is not that mathematical time…it coincides with my impatience, that is to say, with a certain portion of my own duration, which I cannot protract or contract as I like' (p. 10). This means that duration is characterised by the present as an unfolding continuity, as well as the consciousness and memory of the present passing that fold together in relation to one another.

Duration also changes according to the subjective awareness of its transformation in its present. In fact, it is characterised exactly as 'the immediate awareness of the flow of changes that simultaneously constitute differences *and* relationships between particulars' (Parr 2005, p. 79). These relationships refer and defer from one another in a plural and heterogeneous fashion, and includes the very awareness of the state in which movement travels through. Bergson therefore presents a critique against the modern rationalisations of time as a distortion of the qualitative subjective experience of duration.

Even though for Bergson the subjective awareness of time is an important element of duration, subjectivity is not its core element. Rather, duration is a combination of both the awareness of time as a continual unfolding as well as the perpetual bifurcation of the present passing through into the past. Additionally, for Bergson, there is a difference between psychological and physiological time. While both qualities of time strongly affect one another, and they come together to form the whole of duration, their differences also needs to be recognised. In *Bergsonism* Gilles Deleuze has emphasised Bergson's duration as a continual expansion and contraction between the infinite possibilities and finite actualities within a heterogeneity of rhythms of time.

Bergson describes that this potential infinite nature of duration as, 'psychological duration, our duration, is not only one case among others, among an infinity of others, a certain well-defined *tension*, whose very definitiveness seems like a choice between an infinity of possible durations' (Bergson *The Creative Mind*, 1946, p. 218). Each rhythm of duration is therefore relative to each other while also varying in intensity and being in a state of constant change. These rhythms are also always continuous with a variety of intensities, pauses, and folds that come together and part from one another.

This lived subjective time has been considered by Deleuze as a contrast to the way Western society has represented and conceptualised time. To quote Deleuze: 'Bergson's major theses on time are as follows: the past coexists with the present that it has been; the past is preserved in itself, as past in general (non-chronological); at each moment time splits itself into present and past, present that passes and past which is preserved' (Deleuze *Cinema 2,* 1989, p. 82). The overarching whole that encompasses these relationships between past, present, and future involves the continual passage of transformation from one state to another. Duration is therefore not linear in nature, but instead involves a heterogeneity of transformations.

Deleuze's approach to Bergson's duration is significant as he emphasises intuition as a philosophical method for duration. Intuition is to perform a conscious self-awareness of the temporal. It is an apprehension of the real that exists before perception or conception in the present (Bergson *Matter and Memory*, 1911, pp. 13–14). Intuition further positions the act of perception to be another element within and of duration.

This approach highlights the continually folding multiplicity of time that is subject to various flows, intensities, and rhythms. To understand movement via intuition is to engage with a self-awareness of the temporal moment: the exact present moment of movement is negotiated according to both its actual and virtual forms. Movement is evidence of the actualisation of a multiplicity of virtual forms in time, and the awareness of this is engaged through intuition. This is evident when Bergson states that 'our present falls back into the past when we cease to attribute to it an immediate interest' (*The Creative Mind,* p. 179). The self-awareness of time through intuition is a valuable philosophical tool for understanding the nature of perception, as it creates an opening for how perception (and perception of time) varies in intensity. The latter is a significant point to keep in mind during the following analysis of Moholy-Nagy and Gabo, who are also interested in perceiving perception through their kinetic works.

6.3 Moholy-Nagy: New Perceptions of Time and Technological Expansion

In the past *Light Space* Modulator has been interpreted as a pioneering work of avant-garde kinetic sculpture, light architecture, early media art, virtual art, mimetic robot, musical spectacle, and relic from early years at Bauhaus. The artwork has more recently been regarded by Joyce Tsai as an early instrument of the artist's consideration of a total work of art (pp. 227–303), and by Jonathan Crary (2008) as a work of art that paints with light rather than a work of kineticism ('Spinning Histories', p. 42). Additionally Lloyd C Engelbrecht has published an extensive biography on Moholy-Nagy and positions the artist, as an influential 'mentor of modernism' in our art history.

Light Space Modulator has also featured in many art historical frameworks in part because Moholy-Nagy applied himself to a broad range of artistic practices and debates around the role of art in society. The artwork has also been approached as an experimentation around the intersections of technology, science, and art. The artist's wife, Sibyl Moholy-Nagy reflected that the work has been described as a work of lumia art, demonstration of kinesis, theatrical prop, robot, mobile, and even domestic furniture, before it was finally described as a kinetic sculpture. These factors enable the *Light Space Modulator* to appear in a multiplicity of interpretations within contemporary art history.

As Smith argues, there is a tendency for contemporary art institutions to perpetuate modern conceptions of creating, experiencing and interpreting art that effectively remodernises contemporary interpretation. The revival of Moholy-Nagy's artistic career, and in particular *Light Space Modulator* has injected renewed interest in kineticism as an expression of time. More specifically the exhibition of early avant-garde kinetic sculpture perpetuates a specific consciousness of temporality in art. Moholy-Nagy's exploration of movement in art across a variety of media, as I will argue, contemplates the possibility of new technological media to be used to produce new images of temporality, and thereby expand the perceptual field in art.

Moholy-Nagy's early interest in sculptural movement was in many ways concerned with the machine aesthetic as a tool to explore the perception of temporality through mechanical media. This perspective problematises Bergson's critique, because actual movement is used as one element within an entire mediated spectacle of motion rhythms. For Moholy-Nagy technological expansion has the ability to open a range of perceptions and produce new expressions of temporality. What is equally important to the accumulated conceptualisations of movement and time between Gabo, Moholy-Nagy, and Bergson, are the differences between these figures as a means for extending the questions and affects of kinetic dynamism in art.

As with Bergson, Moholy-Nagy was primarily concerned with perceiving continuous movement, and its affect. Moholy-Nagy emphasises that his kinetic sculptures do not create movement from stasis. Instead they are 'carriers of movement' that emerges from an incipient state and the object of movement is therefore subsidiary to the form of movement. For Moholy-Nagy the wider mainstream discussions of the sciences in relation to the arts, 'designated a new dynamic and kinetic existence freed from the static', in which his artwork was used to create new perceptions of time, space and movement (*Vision in Motion*, p. 266). It is these explorations of kinetic sculpture in relation to the perception of time and space by Moholy-Nagy that provides openings for kinetic art to later develop as an aesthetic.

First shown in Paris, 1930 at an exhibition on Deutscher Werkbund at the Salon des Artistes Décorateurs de Paris organised by Walter Gropius, László Moholy-Nagy and Marcel Breuer *Light Space Modulator* consists of three intersecting metal axes, which are fitted with various metal and wood elements comprised of spiralled, perforated, and curved forms hanging from them. The work is clearly modern industrial in its aesthetic; its visible cogs, frames, and a simple mechanical motor—that propels the motion of the axes—the silver steel panels, spirals, and perforated plates that hang off them are notable products of their time. Each of the axes rotates at their individual speed, and also moves with the rotation of the entire sculpture. The structure is mounted to stand at 151 cm high, and is installed with a simple motor to propel its anti-clockwise gyration that completes a rotation every two minutes. As the sculpture moves light projects from beneath the work, then reflected and refracted out into the exhibition space by the hanging axes. This produces a variety of textures of light and shadows, which are cast out onto the walls and space that surrounds the work. Its performance of light projection, reflection, and modulation activates and brings attention to the space around it with an illumination of moving light textures.[1]

The work performs two kinds of movements simultaneously: the mechanical rotation of the axes as they turn, swing, and meet with one another, and the motion of the projected light modulations, textures, and shadows cast onto surrounding areas. When in motion light is projected onto the work from lights fitted within it, to be reflected by the axes and the hanging elements, activating the space around it with a variety of light textures that are reflected off the axes and their hanging parts swinging from the motion. Moholy-Nagy manipulated light, movement, and mechanics to create a sculpture that is intended to transform through time from a static mechanical structure to a weightless body of light and movement. As he explained, 'sculpture is both material volume and its transformation into virtual volume; it has tactile existence but may be changed to visual grasp; from static to kinetic; from mass to space-time relationships' (Moholy-Nagy *Vision in Motion*, p. 237). Depending upon the speed of motion, 'the originally heavy block of material—the solid volume transforms itself into a kind of ethereal extension'

[1]At its initial exhibition in 1930 at the Salon des Artistes Décorateurs de Paris in an exhibition on Deutscher Werkbund, the light textures were prevented from projecting across the room because the work was encased inside a box with a single hole cut out on one of its faces for viewing. However, this was the only time that it was exhibited this way and it has since appeared open in the centre of exhibition spaces, and within theatre stages, to cater for panoramic viewing. See: Moholy-Nagy, *Vision in Motion*, p. 238.

(*Ibid*). This highlights that kinesis, for Moholy-Nagy, was considered as a tool to produce a balance between actual mechanical and optical sensations, to negotiate the materiality of sculpture and its weightlessness as a carrier of movement.

The *Light Space Modulator* is an object that presents motion as a constant transformation of rhythms. The sculpture performs genuine motion, reflects light—and deflects attention—into the space in a rapid, entropic display. The reflections also activate the dimensions of the 'empty' space and display what Moholy-Nagy terms 'virtual volume', which I will address later in the chapter. What is important at this stage is that Moholy-Nagy orchestrates two effects: actual movement and the vibrating virtual volume. Movement is positioned as the fulcrum that connects both actual and virtual images together, forming a whole ensemble of movement. Rather than producing an image of movement that is outside the space and time of the art goer, the multiplicity of movement itself is highlighted through the virtual volume.

In addition to its kinetic movement, Moholy-Nagy created the *Light Space Modulator* to be a machine of mediated movement rhythms through the theatrical, photographic, and cinematic representations of its form and movement. As a machine that produced multiple experiences and representations of movement, this is precisely the productive nature that Moholy-Nagy had referred to in 'Production-Reproduction' (Kostelanetz 1971, pp. 289–290). Rather than presenting mediation as a reduction of movement, he demonstrates that the *Light Space Modulator* performs multiple experiences and mediations of movement, which are produced from it, thereby mobilising the perception of movement in relation to space and time. It is a work that produces forms, perceptions, and expressions of movement and time through mechanical repetition.

Perceiving both the object in movement as well as optical light rhythms are the key dual elements of Moholy-Nagy's work. The sculpture, for example, creates genuine motion, reflecting light into the space in an entropic display of rhythms, repetitions, and collisions of both metal and light. This transformation is based on stasis, movement, modulation, and refraction, each movement creating an entirely new quality of movement that defers from a linear process from movement to acceleration and speed. This orchestrated modulation of light and metal is subtly balanced out with a simple control mechanism that enables the sculpture to be programmed (albeit to a limited extent), where 'the glow-bulbs flash at different places according to a prearranged scheme' (quoted in Passuth 1985, pp. 310–311). Again, there is a dual balance; the sculpture is both chaotic in it movements, as well as controlled.

A critical point of difference between Bergson and Moholy-Nagy's motion studies lies in Bergson's resistance to mechanical movement being used for creating perceptions of duration. As noted above, Bergson argues that mechanical media, film, and photography reduce the heterogeneous and qualitative properties of the experience of time. In contrast, Moholy-Nagy prevents representations of movement from being mistaken as actual movement by building a modality of movement that varies between his sculpture, film, and photography. Each representation and production of movement is used to create and explore new effects of movement.

Additionally, Moholy-Nagy uses mechanical media to provide avenues for contemplating the changing tendencies for how temporality is conceptualised by society. For Moholy-Nagy, kinetic sculpture is one of many ways to explore how visual media carry movement. This is clearly indicated when he says, 'this accounts for the permanent necessity for new experiments. *From this perspective, creative activities are useful only if they produce new, so far unknown relations*' (author's italics) (*Vision in Motion*, pp. 289–290). The *Light Space Modulator* is evidence of Moholy-Nagy's intention to explore perceptions of movement and time across various media, which kinesis indicates. The whole orchestration of movement between the *Light Space Modulator, Lightplay: Black, White and Grey,* and accompanying photographic studies of motion, are direct manifestations of this intent, and point towards technological production—and at times, reproduction—of modern media as a means of creating difference. Therefore, in contrast to Bergson's argument that mechanical representation can only produce a reduction of duration, Moholy-Nagy brought attention to the perceptible differences between mechanical presentation and representation of movement.

This approach to studying light and motion became central to Moholy-Nagy's experiments with sculpture, photography, and film during his time as a teacher at Bauhaus. The dynamic effects of motion in art were already heralded as a key trope of modernity throughout various avant-garde movements in Europe, to which Moholy-Nagy regularly attributed the creations of in his essays and texts (*Vision in Motion*, pp. 198–203). Actual movement and light were useful for creating 'new possibilities of optical and kinetic creation' within art and the greater public arena (quoted in Passuth 1985, pp. 310).

Prior to 1930, sculptural kinetic experiments had appeared in the wider artistic arena within Europe, and Moholy-Nagy's *Light Space Modulator* is often grouped with these earlier avant-garde productions in contemporary art history (Burnham, *Beyond Modern Sculpture,* 1968, pp. 235–238). Aside from Gabo's *Standing Wave; Kinetic Construction* (1920), which was a source of inspiration for Moholy-Nagy (*Vision in Motion*, pp. 226, 238), constructivists had previously worked in participatory sculptural movement, and Marcel Duchamp and Man Ray had collaborated on the optical illusions produced by *Rotary Glass Plate, Precision Optics* (1920). While these works are often framed by Western art historical arguments as avant-garde efforts for collapsing the division between art and life, unlike Moholy-Nagy's *Light Space Modulator* they were predominantly regarded by the artists as time-space experiments in preparation for following works of art.

While the *Light Space Modulator* was not the first sculpture to perform actual motion, works such as Marcel Duchamp's *Bicycle Wheel* were purely considered as an experiment with motion (Duchamp 1973, pp. 141–142). *Light Space Modulator* is considered by historians such as Joyce Tsai as the earliest form of kinesis that exists as a sculpture in its own right (Tsai 2011, pp. 227–303). It is also one of the earlier known sculptures to present and modulate its own movement for multiple environments and exhibition spaces in contemporary art history. Moholy-Nagy's use of modulation, refraction, and illusion with mechanical media is directly in tune with the artist's intention to explore the creative and productive uses of industrial media rather than their reproductive functions.

Sculpture in movement played a pivotal role in Moholy-Nagy's frustration with film and photography at the time. In 1932 Moholy-Nagy considered film to still be 'governed by conceptions derived from traditional studio painting', rather than as a medium that achieves what others cannot. Moholy-Nagy particularly stressed that, 'motion is still so primitively handled in the majority of films' (quoted in Kostelanetz 1971, p. 131). The orchestration of actual movement in sculpture and the study and recomposition of motion in film enabled Moholy-Nagy to explore the differences between action motion and its mediation, as well as the possibility of 'painting with light' and with movement. Working with non-figural light as a medium itself enabled Moholy-Nagy to prevent distinctions between the original light source and the modulated image of light. The hanging and rotating fixtures of the piece enable the reflections to intersect with one other, providing both fractions and convergences of light that ensure that the viewer cannot distinguish a hierarchy between images of light, their motions, refractions, and convergences.

The *Light Space Modulator* was not created solely as a moving sculpture for exhibition, but also as a machine for creating multiple representations of movement in film (titled, *Light Play: Black, White and Gray*), photography, and creating moving background settings in theatre (Tsai 2011, pp. 227–303). As a machine that produced multiple experiences and representations of movement, this is precisely the productive nature that Moholy-Nagy had referred to in 'Production-Reproduction'. Rather than considering mechanical mediation as a reduction of movement, Moholy-Nagy demonstrates that through the *Light Space Modulator* multiple experiences and mediations of movement are produced. It is a work that produces forms, perceptions, and expressions of movement and time through the technology of mechanical reproduction.

The film *Light Play: Black, White and Gray* demonstrates Moholy-Nagy's interest in using film to create new effects of movement and time through the cut of mobile sections. The film's only surviving sequence shows a variety of light sources, and focuses on their intensity and texture of light, each differing from the next. As Moholy-Nagy's script describes, '[d]istortion of reflections. Pendulum. Blinding moving light flashes. Revolving spiral, reappearing, again and again. Rotation increases; all concrete shapes dissolve in light' (quoted in Kostelanetz 1971, pp. 149–150). Each scene within the montage is a visual break down of a texture of light, and it is a linear rendition of what is orchestrated by the kineticism in Moholy-Nagy's sculpture.

Each scene in the film produces an image of time that has the ability to mobilise the perceptions of space and time within this media. Attention is also paced out to focus on producing and enframing new perceptions that the sculpture produces. This is precisely the difference that Sibyl Moholy-Nagy, his partner at the time, also noted when she described the first meeting with the Light Space Modulator:

> In the centre of a workshop stood a construction – half sculpture and half machine – a combination of chromium, glass, wire, and rods, in which I recognized the forms of the light-display film. As it turned slowly, invisible lights flared up and turned off, producing gigantic shadows on the wall and the ceiling…and I suddenly saw the difference between concept and reality (Kostelanetz 1971, p. 147).

To highlight the perceptual differences between the film and kinetic sculpture separates the *Light Space Modulator* from experiments purely concerned with objects in kinesis. This is why Moholy-Nagy's experimentation with movement does not stand alone solely with the *Light Space Modulator* but, rather, a variety of motion rhythms, produced through film, photography, and the sculpture, demonstrates the perceptual differences and unique relationship to movement each medium has in relation to one another.

The artist's notes published posthumously in *Vision in Motion* shows his interest in a moving object transforming in time, inhabiting new rhythms, natures, and structures: 'sculpture…has tactile existence but may be changed to visual grasp: from static to kinetic; from mass to space-time relationship' (129). Both the gradual transformation through time, as well as the change to modulation and representation of the sculpture through photography and film are included within this expression of temporality. Moholy-Nagy's open-ended view of how the sculpture was to be received indicates that there is not only one way to experience the works of art, but that the modulations of light that are produced are multiple.

It is here that a critical point of difference exists between Bergson's duration and Moholy-Nagy's motion studies. Bergson expressed reservation towards mechanical reproduction of movement being used for creating images of duration. His argument centres on film and photography reducing the heterogeneous and qualitative properties of the experience of time, while Moholy-Nagy prevents representations of movement from being mistaken as actual movement by building a modality of variation between his sculpture, film, and photography. This affords Moholy-Nagy the ability to be attuned to each mediation movement as its own experience of duration. Movement is carried by the sculpture in duration, performing rhythms of movement reflections, refractions, and modulations. It is also designed to move between images in his photographic studies and is presented as mobile cuts of movement in film, enabling a process of mediation to produce new sensations and perceptions of motion.

In his influential essay, 'Production – Reproduction', published in *Der Strum* (1923), Moholy-Nagy described his main motivations for interdisciplinary artistic production; experimenting with the technological expansions of the industrial revolution provides a means for producing new stimuli, rather than merely reproducing repetitive experiences. He proclaimed, 'man will be most perfect in his own time if the functional apparatuses of which he is composed…are conscious and trained to the limit of their capacity. Art actually performs such a training' (Passuth 1985, pp. 289–290). This approach echoes the teachings of the Bauhaus (where he was teaching at the time), that recognise that the creation and experience of art in everyday life enables humanity to not only live harmoniously, but also in a higher more complex state.

Even though Moholy-Nagy claimed to be interested in the potential for all artistic media to open 'all the channels of intuition' (Passuth 1985, pp. 289–290), he explicitly strove for the exploration of new industrial media in art. In 'Production – Reproduction' Moholy-Nagy explicitly sought the development of total art through machinic creative production, rather than considering industrial media as a means for repetitive reproduction. Creative production, rather than reproduction, is approached by Moholy-Nagy as a key tenant of avant-garde experimentation as, 'since it is primarily production (productive creation) that serves human construction, we must strive to turn the apparatuses (instruments), used so far only for reproductive purposes, into ones that can be used for productive purposes as well' (289). Considering this, *The Light Space Modulator* participates in an array of experiments with movement, mediation, and modulation as a demonstration of productive creation on multiple levels. Each context of the sculpture plays with the form of each medium inherent to its own capacities, and expressing its own rhythms of motion. Considering 'Production – Reproduction', the *Light Space Modulator* is further evidence of Moholy-Nagy's intention to expand perceptions of movement and time through modern media, within which kinesis is integral to his approach. While actual motion is performed by the sculpture, the transformation, variety, and creative production of this actual movement is of equal importance in the quest to explore the limits of human perception and sensation.

In this regard, technological production is a key practice for exploring the limits of human perception and producing experiences through spectacles of the present. The whole orchestration of movement between the *Light Space Modulator*, the film *Lightplay: Black, White and Grey,* and accompanying photographic studies of motion are direct manifestations of this intent. Through this Moholy-Nagy understands technological production (and at times, reproduction) as a means for creating difference, rather than colonising perceptual fields with technocratic similitude.

Rather than perpetuating the three theses of modern movement as proposed by Bergson, Moholy-Nagy sought the perceptible differences between each mediation, and the temporal play that can be engaged through them, with the intention of forming new expressions of time through industrial media. The experience of time when presented or represented, even when it is divided into distorted recompositions, serialised instances, or modulated by projection and reflection, has the possibility to bring more than three expressions of movement within the whole duration. Western society, Moholy-Nagy shows, is receptive to new conceptualisations of movement through modern industrial mechanisms.

6.4 Blurring Time with Virtual Volumes

In ways differently to Moholy-Nagy's *Light Space Modulator*, Gabo's sculpture *Kinetic Construction; Standing Wave* (1920) problematises the relationship between the presentation, representation, and perception of movement in several ways. Gabo's sculpture was created, as was often referred to by the artist, as an instructional device to demonstrate to his students how to explore the potential for 'kinetic rhythms' to transform the static form of sculpture to an illusory volume. *Kinetic Construction* is a small and simple work that consists of a vertical rod that stands out from its mount (standing at 616 × 241 × 190 mm), within which lies a small motor that propels the rod into motion. The combination of speed of movement, flexibility, and length of the rod creates an appearance of a semi-opaque three-dimensional vibrating crescent, or as Gabo describes, a 'virtual volume' (Kirby 1969, p. 251). This volume expands and pulses according to the stress of the rod before it returns to rest, giving the appearance of a static volume, even though the appearance of movement is evident. While the standing wave is a three-dimensional movement, reconstructions such as those at the Berlinische Galerie in Berlin exhibit the work encased inside a wall with a glass window, which emphasises the optical qualities of the piece.

As with Bergson, Gabo's *Realistic Manifesto* and *Kinetic Construction*, emphasises time as a constant and concrete element defined by qualitative change. Gabo's decision to make the rod propel into motion, and then return to stasis rather than remaining in perpetual motion, may have in some ways been a technological consideration. Despite this possibility, the movement from stasis to motion creates a unique effect: the work performs a qualitative transformation through time from a stationary rod into a vibrating wave of movement, in time with the viewer. Gabo's virtual volume reveals that the perception of time can be manipulated through actual movement, rather than static representation, without distorting or simplifying the duration as a transformation. This is what Michael Kirby has described as the 'transitional mode of kinetic perception in kinetic sculpture: changing perception occurs in time with the viewer, whose perception, moves with the movement of the sculpture' (p. 253). For Gabo, this is the key function of the virtual volume. Rather than creating illusions *of* movement, this optical image is created *from* actual movement.

For Gabo, movement is a flow of constant change, and this includes the actual movement of the sculpture, as well as the optical images it produces. This is a key point of difference between Bergson's duration and Gabo's kineticism. While Bergson is open to the movement of the projector, camera, and operator in film and photography as a form of duration, he remains adamant that duration cannot be produced as a representation (*Creative Evolution* pp. 320–323). In contrast to Bergson, Gabo's *Kinetic Construction* produces actual and optical experiences of movement that are mutually exclusive, and yet defer to one another. Gabo presents both a presentation and a representation through movement, rather than representational distortion.

While for Bergson mediation, modulation, and illusion produce simplified representations of the nature of duration, Gabo uses the limits of human perception to produce an image that is an accumulation of tenses. The volume presents perception as something that is malleable, and subject to manipulation in relation to perceiving time. The blur of the standing wave produces an image that is an amalgamation of both present and past motions that stretch together into one blurred impression. Understanding the perceptual limits of movement is therefore to acknowledge the perceptual limits of time, with the aid of modern machinery. This resonates with Bergson, who considered that 'what is real is the continual *change of* form: *form is only a snapshot view of a transition*…our perception manages to solidify into discontinuous images the fluid continuity of the real' (*Creative Evolution,* p. 12). However, just as importantly, unlike Bergson, Gabo develops the movement of the rod as a continual change in order to produce an optical distortion of a new form that waves and vibrates, in a semi-opaque form.

Gabo's *Realistic Manifesto* (1920) was exhibited by being read aloud by Gabo and his brother Antoine Pevsner. It is arguably the earliest piece of literature that explicitly refers to kinetic motion as a tool for criticising traditional modes of representation. The manifesto concentrated on creating a new order of art that drew upon actual movement, time, and space rather than depictions of them. Frequently described by Gabo and his peers as a key early Constructivist statement, his manifesto claimed, 'building the future, building a new society and culture, building a new art in accordance with the form as of space and time on which life is built' (quoted in Hammer and Lodder 2000a, b, p. 22). Like many avant-garde manifestos of the time, *Realistic Manifesto* collapsed the distinction between art and life by championing the chaos and complexity of real life, movement, space, and time. Composing real time and actual movement in sculpture prevents metaphorical and fictional expressions. By regarding them as independent concrete forces, and composing actual experiences of them, Gabo concentrated on the 'perceptions' of the forms of space and time, as well as how these perceptions change, and that representation is a distortion of space and time. Gabo and Pevsner state that bringing such changes into being are the 'sole aim' of their art in the manifesto.

Exhibited as a poster during the *Open Air* exhibition with Hustav Kucis on Tverskoie Boulevard in Moscow, *Realistic Manifesto* criticised the fictional and metaphorical functions of representation used by modern artists to connote speed, rather than celebrating actual mechanical and modern movement. This was a key mechanism for motivating and ensuring the futurity of socialism throughout the nation; an artistic practice that, so they claimed, does not lie to its people. In order to achieve this, the manifesto states, real-time movement in art propels its viewers to bring their attention to the present, while representations of movement provide images of the past: '[t]he attempts of the Cubists and the Futurists to lift the visual arts from the bogs of the past have led to only new delusions…We assert that the shouts about the future are for us the same as the tears about the past: a renovated day-dream of the romantics…Today we take the present' (93). While representation provokes perceptions of the past, Gabo and Pevsner encourage the perception of perceiving in the temporal present as a means for attaining attention of the present in the present.

While Gabo was also intent on understanding time as a perception of change, or in his words, 'concrete movement which changes before the eyes' (93), *Kinetic Construction* demonstrates Gabo's interest in bringing not only real time into consciousness but, also how the perceptions of it change. Real movement of actual matter was the key tool to making time 'active and perceivable', rather than creating significations that distorts its rhythms (Hammer and Lodder 2000a, b, p. 42).

Therefore, what is unique to Gabo's approach to actual movement in relation to Moholy-Nagy and Bergson is how he understood mechanical motion and repetition to persuade time-space perceptions. The movement of *Kinetic Construction* draws attention to the heterogeneous nature of mechanical media in motion that holds both discrete and continuous properties. The sculpture is both rapidly moving while giving an impression of a vibrating form, it is both actual and illusory, and produces perceptions of the present and past within the one transformation. At the same time, the work is also binary and discrete; it is a self-contained work that is either in motion, or not. Each performance of movement is then an interval of movement that returns to extended periods of stasis.

Gabo, like Moholy-Nagy, also saw mechanical production as a means for creating a new order of art, with new perceptions and sensations that were not previously activated prior to the technological expansion of the industrial revolution. Perception is open to illusion, as demonstrated by *Kinetic Construction*, which transforms in time to produce an optical volume that is also an alternative impression to the actual moving rod. In *Realist Manifesto* Gabo states that his interest in the 'basic forms of our perception of real time' (Danchev 2011, p. 93), suggests that *Kinetic Construction* is as much a work about actual motion as it is about the perception of and within temporality. To bring the present into consciousness through actual motion is to also create an image of temporality during the artistic experience. Focusing on the present is best explored through artworks that exactly work within this temporality in actual form, which were already in existence in the arts of music, theatre, and dance. It is in this respect that movement of mechanical media was used to open new expressions of temporality that had otherwise been unexplored by European artists at the time who had largely concentrated on images of past and future.

6.5 Accumulating Temporalities

The construction of actual movement and virtual volumes in motion is evidence of Gabo's resistance to traditional modes of movement in modernity. Unlike painting, film, and photography, Gabo considered kinesis as the primary means to engage the viewer in the present temporality. The virtual volume therefore exists as an articulation of time as a transformation in the present. The volume also produces an impression of time that consists of various folds of temporalities. While the rod moves in real time, the virtual volume is produced because the human eye cannot accurately trace and process its movement. In that sense past impressions of its movement are preserved during perception, which produces the virtual volume. The volume is therefore an accumulation of perceived temporalities. This indicates that the virtual volume is a technique for artists working with kinesis to use movement to create new conceptions of temporality. Rather than considering time as a linear state, Gabo's virtual volume articulates time as an accumulation of states, as well as a continuous transformation. Mehanical movement, repetition, and the effect of the virtual volume is used not to distort but, to encourage new images of duration.

The virtual volume is another effect that connects kinesis with new perceptions of temporality with kinetic sculpture. The earliest reference to virtual volumes is Gabo's *Standing Wave,* which explored the edge between the transformation from movement to static form to optical volume. During this transformation viewers perceive-with the work: that is, the volume is not an image of transformation but, an experience of transformation in real-time. Each pulse, rhythm, and repetition is also felt with the optical illusion of an alternative volume that exists in the same temporality. The transformation from a static rod to a virtual volume creates an appearance of weightlessness. This is because the volume pulses and oscillates seemingly unattached to its base as the rod bends and repeats its motions. It is an effect that is actual in its temporality but illusory in its form because its optical appearance (the volume) is different to its actual form (a straight rod).

It is important to note that Gabo's use of the term 'virtual' here differs from Bergson's understanding of the term. For Bergson, the virtual can be considered as the infinite state of forms that becomes actualised in the present in its continual state of becoming, and yet does not temporally coincide with the actualised present. Despite seeming like opposite conditions, the virtual and the actual are equally real. To consider the virtual through the role of affect in art is a useful example: While affects are exactly what a work of art produces, they are only actualised through perception. For Bergson, the affects that are not produced lie dormant in a realm of chaotic potential until they are actualised.

Unlike Bergson's use of the term 'the virtual' Gabo's virtual volume in *Kinetic Construction* is an optical sensation in actual time. It is an optical effect of rapid movement. For Gabo, 'the standing waves had attracted my attention since my student days, in particular the facet that when you look at a standing wave, the image become three-dimensional…I decided to construct a standing wave which would be vibrating on one fixed point and rigid enough to indeed be a '*standing* wave" (quoted in Hammer and Lodder, *Constructing Modernity*, 2000a, p. 69). Gabo straddles a line between what is optically actual through appearance but what is not actual matter. The material rod and its virtual volume exist simultaneously within the same temporality, which gives a point of indiscernibility between the form volume and actual rod. To see a form in motion that is an alternative to its actual composite suggests Gabo's approach to virtuality as a real entity providing the object in motion with a multiplicity.

Kinetic Construction was Gabo's first and only attempt to orchestrate motion in this manner, and had been referred to by the artist as an early example for future, more complicated kinetic rhythms (Hammer and Lodder, *Gabo on Gabo*, 2000b, p. 113). This short lived experimentation with kinetic dynamism is due to the artist's belief at the time that technology would not advance far enough in his lifetime to be able to explore other rhythms of actual movement; the possibility and experimentation of kinetic dynamism was considered by Gabo as a 'task for future generations'.

Years after Gabo's *Kinetic Construction* Moholy-Nagy drew inspiration from Gabo's work and considered *Light Space Modulator* and his photographic studies of virtual volumes as spectacles of the modern technological expansion. In *Vision in Motion* Moholy-Nagy explained both sculptural and photographic media as explorations of the form of movement as well as the affect of movement in real-time. His blurred photographs are static representations of virtual volumes by showing the repetitive motions or movement that extend and pass through over time within a single frame. Rather than capturing and isolating instances, they, like the experience of viewing *Kinetic Construction,* express movement through time that has been synthesised together. Unlike other practices of photography at the time, which were used to render perceptible what is ordinarily imperceptible.

Moholy-Nagy's photos explored the new representations of movement and time specific to the medium. Their virtual volumes are often compositions of urban scapes, neon flashing signs, long exposures of street traffic at night, carnivals, and spinning toys. They demonstrate that the modulation of time and space is already within the modern everyday. For Moholy-Nagy virtual volumes were not only produced by kinetic sculpture and photography but, were laced within spectacles of modern urban life. Everyday uses of artificial light were key sources for experiencing virtual volumes outside art. It is partly for this reason that light was considered a democratic artistic medium, because it can be created and consumed in everyday experiences in modern urban life.

Alternatively, another work by Moholy-Nagy, *Kinetic Sculpture* (1930–1936), is an apparatus that unlike the *Light Space Modulator* and *Standing Wave; Kinetic Construction,* is not propelled by an industrial motor, but by the viewer's participation. The work consists of two square cage-like forms, made from glass tubing filled with mercury, which protrude out from a steel plate and can be spun in rotation. The speed of their motion, as well as the reflections cast from the glass and mercury, create visual effects of virtual volumes similar to Gabo's *Kinetic Construction.* Again, the work transforms from static material to the weightless virtual volumes through participation and motion.

The virtual volumes developed by each of these two artists are explicit attempts to draw out and expand perceptions of the present. What is specific to Moholy-Nagy's photographic study of virtual volumes is that they are singular images of the past and present in one temporal frame. This encourages a perceived image of the present that is expanded, stretched out, to portray a crystalised continual transformation. Whereas the virtual volumes created in *Light Space Modulator, Kinetic Sculpture,* and *Kinetic Construction* are produced in real-time with the viewer's perception, rather than presenting a whole synthesised image of time continually unfolding.

The transformation from heavy materials to an appearance of weightlessness is also integral to Moholy-Nagy's appropriation of the virtual volume. As he describes, the sculpture 'transforms itself into spheric extension, becomes completely without mass and heaviness' (*The New Vision*, p. 129). Both material form and the weightless volume simultaneously exist in the one form but optically transform to express a dynamic of 'volume relationships'. The sculpture presents a modulation of movement in its performance but, also a modulation of time through the virtual volume. Gabo and Moholy-Nagy's virtual volumes attempt to give an appearance of matter transforming into light and movement.

We also feel the very edge between the rhythms of movement in kinetic sculpture: the transition from weighty material *into* virtual volumes, or the movement of light as it is modulated, refracted, and projected. In these movements of qualitative transformation there is both the experience of movement carried by the work, as well as a sense of the movements that are not actualised. These works not only make visible the invisible, in the sense that they are virtualisations (in a Bergsonian sense) from the chaotic field of potential forms but we also feel their intensities through the composition of actual movement. Perception of any artwork — static or moving, comes in drabs, fragments, rhythms, and holes. Gabo and Moholy-Nagy's use of repetitive motions further makes it clear that there is difference in each repetition of rhythm, that is, the production of affect through duration.

Both Gabo and Moholy-Nagy explored the felt hesitation of change in the temporal present and the nature of this temporality as a productive event. Rhythm, repetition, and speed create a new image—the virtual volume—and draw attention to the present temporality. The elements of speed, repetition, and rhythm, also create a visual image for the viewer, which is an optical effect. Although mechanically repeatable each artist focuses on the transformation from one move or action to another as a state of becoming as the temporal present unfolds.

Likewise, Moholy-Nagy's *Light Space Modulator* was intended to draw attention away from the mechanical component of the sculpture, and dissolve into an orchestration of modulated light textures. Using light to give the appearance of visually dissolving matter was a key effect for these artists concerned with transforming matter into weightless optical sensations and virtual volumes. For instance, the use of light in *Light Space Modulator* and *Kinetic Construction, Standing Wave* permits the sculptures to not only transform in their own rhythms of duration but also perform a variety of movement intensities. Rather than considering intensity according to speed, both artists consider the intensity during perception closer to the rhythm of change in movement.

While for Bergson intensity is 'the image of a present contraction and consequently a future expansion' (*Time and Free Will*, p. 4), the very awareness of this edge between contraction and expansion could be an effect of assembling movement in real time together. This refers to a specific event or rupture that creates a qualitative difference in the art-goer, which occurs during an encounter with art. Deleuze (1990) extends this notion further by stating that speed forms a role in enhancing the affect of intensity (*The Logic of Sense,* p. 147). From this regard intensity resides at the interface of affect; a fuelling force even.

Just as virtual volumes could be found and captured in urban spaces, kinetic dynamism for both Moholy-Nagy and Gabo was never intended strictly for the practice of sculpture. Both artists pursued kinesis as an effective form of harnessing attention of temporality that could extend through to experiences of theatre, architecture, and public art spectacles in sites of leisure, arts, and entertainment.

Moholy-Nagy specifically commented on this intention a number of times where he glorified the consumption of real movement in public urban industrial spaces. He reflected on this intention for further possibilities of consuming and orchestrating kinesis in the public domain in the future, when technological advancements had been made:

> I dreamed of light-apparatus, which might be controlled either by hand or by an automatic mechanism by means of which it would be possible to produce visions of light, in the air, in large rooms, on screens of unusual nature, on fog, vapour and clouds (quoted in Kostelanetz 1971, p. 156).

Moholy-Nagy's utopian aspirations envisioned mechanical mediations of movement to be ubiquitous, immersive, and potentially a mass public consumption of art and entertainment. The impermanence, weightless, and immaterial nature of light positioned lumia is also an effective medium for reaching large crowds without the permanence of sculpture and architecture. Light and movement in the public domain had already often been utilised as a means to commemorate human accomplishments, political power, or as gateways to further innovation and consumption of technology.

Moholy-Nagy's intention for kinesis in art was directed at the potential for whole architectural constructions to reconfigure subjective perceptions of space, time, and movement. This is specifically addressed in his (unrealised) project, *The Construction Scheme of the Kinetic Constructive System* (1922), a cylindrical shaped building used as a recreational multi-purpose space structured with multiple layers of spiralling ramps. For this kinetic construction, the movement that is 'carried' is performed not by the apparatus, but the people within it. The inhabitants are instead performers of a spectacle of leisure. Moholy-Nagy dreamed that:

> we can imagine the play of light in community festivals of coming generations. From airplanes and airships they will be able to enjoy the spectacle of gigantic expanses of illumination movement and transformation of lighted areas, which will provide new experiences and open up new joy in life (quoted in Passuth 1985, p. 333).

Moholy-Nagy predicted that kinetic art would be effective for public, or at least communal enjoyment. Additionally, experiencing virtual volumes in urban landscapes such as advertising signs, evening traffic, fireworks and profile searchlights, reveals the virtual volume as an element of the modern everyday. The volumes are therefore, for Moholy-Nagy, also an articulation of the experience of urban temporality and the attention that it demands. He described them as, 'sculpture = material + mass relationships' and considered public activity as catalysts for these mass relationships ('A New Vision', p. 129). Moholy-Nagy's unrealised projects using motion and light in open areas, or large rooms with multiple projectors performing an entropy of movement, are an explicit desire for creating an aesthetic that harnesses the perceptual edge between what is moving, how that movement will transform in the future, and how the public also transforms within this spectacle.

If, for Bergson, intensity points to both the swell and expansion of the present that extends to the future (*Time and Free Will*, p. 3), then public, theatrical, and architectural performances of kinesis expand on the temporal present, and can do so with a state of intensity. The virtual volumes transform material sculpture to become both weightless and actual in the present, and also activates the space of the public installation as a specific temporal present. They not only happen in real time but also position the present as constantly unfolding, expanding, and passing, because the volume is a visual accumulation of temporalities. The space activated by the lights in Gabo and Moholy-Nagy's projects is both virtual and actual, both unfolding and refracting in the actual present.

For Gabo, the production of real movement and time enabled the artist to express and critique the unfolding present as an accumulation of multiple temporal modes. Moholy-Nagy's appropriation of Gabo's virtual volume articulated this through his photographic motion studies, and also emphasised the virtual volume as a means for sculpture to simultaneously be mechanical and weightless. The virtual volume plays an integral role throughout Gabo and Moholy-Nagy's experiments with kinetic movement in their sculptures. Not only does the virtual volume produce appearances of a weightless object that oscillates between present and past but they also contribute to a perception of time as a transformation that moves with the viewer's perception through time from mass to weightlessness. The sculptures that I have highlighted in this chapter re characterised as being tactile and optical, material and weightless, virtual and actual, and blur the impressions of present and past actions. Many of the writings by these artists indicate an interest in the forms that mechanical kinesis can bring, as well as the unseen but felt ways that kinesis can affect their viewers. In this light, virtual volumes and kinetic movement exist in time, but offer alternative visual perceptions of temporality—they offer to produce optical illusions which are visually true, but alternative to their actual forms.

6.6 Pioneering a Kinetic a Consciousness of Time

Throughout this chapter I have approached Moholy-Nagy, Gabo, and Bergson as three figures whom, when considered together, form a peculiar consciousness of time. This consciousness is geared towards the perception of universal (duration) and subjective (psychological) time that are continual transformations. Bergson, Moholy-Nagy, and Gabo also consider the role of mechanical movement and representation to articulate duration differently. I have highlighted Bergson's criticism for the way in which Western representations of movement distort the actual experience and perception of time. For Bergson, the gap between representation and the real was more than a problem of misrepresentation. Repetitive reproductions of these distortions led to a common architecture for understanding time in modernity that distorts the subjective experience of time and confuses mediated representations of movement with movement itself.

Moholy-Nagy and Gabo's study of kinesis have been used to problematise Bergson's regard for modern mechanical media in a number of ways. These artists both consider kinesis in art as a tool for drawing attention to the present temporality in an attempt to explore the role and function of time in art. Moholy-Nagy's *Light Space Modulator* is highlighted as a machine that performs movement as a kinetic sculpture, but also a machine that Moholy-Nagy mediates through film and photography. The *Light Space Modulator* is therefore a machine that moves and is also a production of movement, mediations, and rhythms that articulates Moholy-Nagy's interest in the creative production of industrial media. Rather than distorting time, the study of movement, modulation, and mediation through kinesis, film, and photography is Moholy-Nagy's attempt to explore different ways to articulate movement, space, and time.

While Bergson was wary of mechanical representation and reproduction, this chapter has highlighted Moholy-Nagy's interest in mechanical kinesis and representation as a means for producing new perceptions and experiences of movement in relation to space and time. Each perception, mediation, and modulation is not considered as a reduction of duration, as Bergson expressed but, a new perception of movement within the viewer's duration. In this respect, mediation has the ability to assemble new percepts and affects to experiences of art. Through *Light Space Modulator* Moholy-Nagy uses kinesis to experiment with mechanical movement as an arrangement of multiplicity. The mediation and representation of the sculpture of movement from Moholy-Nagy's perspective does not produce a reduction of movement but, instead, additional perceptions of motion rhythms. The creative productive abilities of mechanical motion open opportunities for new perceptions of, and relationships to, temporality that can be orchestrated in relation to actual duration.

Virtual volumes produced by Gabo, and later appropriated by Moholy-Nagy, further emphasise the ability for kinetic sculpture and photographic media to articulate the present as an accumulation of temporalities as they unfold. Viewing *Kinetic Construction* is an experience of movement that synthesises perceptions of the present and past together. Rather than attempting to spatialise time, virtual volumes are produced to highlight movement that transforms in real-time duration, with the view to creating a new understanding of time in art. Moholy-Nagy and Gabo's creation of virtual volumes through kinetic sculpture is also significant because they regard the virtual as a term that depicts materiality and immateriality simultaneously: these sculptures are tactile while optical, material and weightless, virtual and actual, and blur the perception of present and past together. In this light, virtual volumes and kinetic movement exist in time, offering alternative visual perceptions of temporality, and producing optical illusions that are visually true but also, alternative to their actual forms.

For Moholy-Nagy and Gabo the virtual volume is an optical effect that is a product of the movement of modern mechanical form. In this regard Moholy-Nagy describes sculpture as 'both material volume and its transformation into virtual volume; it has tactile existence but may be changed to visual grasp; from static to kinetic; from mass to space-time relationships' (*The New Vision,* p. 129). In addition to considering early avant-garde kinetic art as a primitive form of digital and virtual art, this consideration for movement to allude to and create the virtual also creates an opening for considering the articulation of duration, and the virtual in contemporary installation. The next chapter addresses Olafur Eliasson's *Your Negotiable Panorama* (2009) as a continued desire to articulate the present as ancrystallisation of temporalities, and the use of motion to guide a perceptual edge between actual movement and its emergence from incipience.

References

Moholy (2012) An education of the senses, Lyola University Museum of Art. http://www.luc.edu/luma/news/011910_moholy.html. Accessed 15 November 2012

Bergson H (1913) Creative evolution, trans. A. Mitchell, Macmillan, London

Bergson H (1911) Matter and memory, trans. N. M. Paul and W. S, Palmer, Macmillan, New York

Bergson H (1910) Time and free will; an essay on the immediate data of consciousness, trans. Pogson, S. Sonnenchein & co., London, F.L

Bergson H (1946) The creative mind, trans. M. L Andison, Philosophical Library, New York

Burnham J (1968) Beyond modern sculpture: the effects of science and technology on the sculpture of this century. G. Braziller, New York

Crary J (2008) Spinning histories: displacement and reversal in the arts of perception. In: Christov-Bakargiev C, Blazwick I, Crary J, Harrison C (eds) 2008 Sydney biennale: revolutions- forms that turn. Thames & Hudson, Sydney

Danchev A (ed) (2011) 100 Artists' manifestos; From the futurists to the stuckists. Penguin, London

Deleuze G (1991) Bergsonism, trans. H. Tomlinson & B. Habberjam, Zone Books, New York

Deleuze G 1989 *Cinema 2: the time-image*, trans. H. Tomlinson, R. Galeta, University of Minnesota Press, Minneapolis

Deleuze G (1990) The logic of sense, trans. M. Lester and C. Stivale, Columbia University Press, New York

Duchamp M (1973) The writings of Marcel Duchamp. De Capo Press, New York

Hammer M, Lodder C (2000a) Constructing modernity: the art and career of Naum Gabo'. Yale University, New Haven

Hammer M, Lodder C (eds) (2000b) Gabo on Gabo; texts and interviews. Artist Bookworks, Sussex

Kirby M (1969) The art of time: essays on the avant-garde. E.P Dutton & Co, New York

Kostelanetz R (1971) Moholy-Nagy. Allen Lane The Penguin Press, London

Moholy-Nagy L (1965) Vision in motion, 2nd edn. P. Theobald, Chicago

Nisbet P Light prop shines again. Harvard Gaz Archi. http://news.harvard.edu/gazette/story/2007/07/light-prop-shines-again/. Accessed 5 August 2012

Parr A (ed) (2005) The Deleuze Dictionary. Edinburgh University Press, Edinburgh

Passuth K (1985). Moholy-Nagy, Thames and Hudson, London

Popper F (2007) From technological to virtual art. The MIT Press, Cambridge, Massachusetts

Ross C (2012) The past is the present; it's the future too; the temporal turn in contemporary art. Continuum International, London

Shanken E (2009) Art and electronic media. Phaidon Press, New York

Solnit R (2003) Motion studies: time, space and Eadweard Muybridge. Bloomsbury, London

Tsai J (2011) 'The sorcerer's apprentice: lászló moholy-nagy and his light prop for an electrical stage', the aesthetics of the total artwork: on borders and fragments. John Hopkins University Press, Baltimore

Chapter 7
Your Negotiable Panorama: The Seeping Edges of Perceiving Yourself Perceiving

Light is not the opposite of materiality.

(Bal 2007 p. 169)

Abstract This chapter focuses on a contemporary installation by Olafur Eliasson titled *Your Negotiable Panorama* as a useful artwork for rethinking the relationship between the movement of mechanical media and perceptions of the virtual. This chapter is focused on movement that is used to perceive the virtual as a means to produce material and immaterial effects, and to contribute to a consciousness of time that resonates with Moholy-Nagy, Gabo, and Bergson's understandings of the virtual. My approach cleaves new possibilities for considering a history of contemporary kinetic art history that does not move from mechanical kinesis towards immaterial digital art, as suggested by Frank Popper. This chapter draws from Henri Bergson's understanding of the virtual as a part of a chaotic field that becomes actualised in the present to argue that perceiving movement in works such as *Your Negotiable Panorama* can act as a visual gauge for negotiating the perception of the actual from the virtual, and for contemplating the crystalline images that arise when the distinction between the actual and virtual are indiscernible from one another.

As demonstrated in the previous chapter, Naum Gabo and László Moholy-Nagy used virtual volumes to consider the simultaneously material and immaterial effects of movement. However, since their kinetic experiments in the 1920s 'virtuality' and 'the virtual' have developed as ambiguous terms in the field of digital art. Popper and Nechvatal, like others, has considered virtual art as a tendency that 'leads continuously from materialized' work towards immateriality, which involves 'the humanizing of technology through interactivity and neocommunicability as well as sensory immersion and multisensoriality' (2004, pp. 65–66). Popper argues in *From Technological to Virtual Art*, that virtual art stems from early avant-garde kinetic artists such as Gabo and Moholy-Nagy and their experiments with virtual volumes. For Popper, these artists are chronologically and conceptually rudimentary precursors of virtual art because they developed the optical effects of weightlessness in their work (2007, pp. 17–18). Furthermore, Popper considers that the study of virtual art necessitates a shift away from the study of motion, towards a cybernetic systems based vocabulary. While this position facilitates and encourages

© Springer Nature Singapore Pte Ltd. 2017

C. Chau, *Movement, Time, Technology, and Art*, Springer Series on Cultural Computing, DOI 10.1007/978-981-10-4705-3_7

a rigorous study of digital and virtual art, it does so by suggesting that actual movement in art is not effective for discussing the representation or allusion to the virtual.

However, as I will argue in this chapter, there is a continued tendency in art to use movement in a similar manner to Moholy-Nagy and Gabo's experiments with movement to create virtual volumes. Rather than considering kinetic virtual volumes as an early form of postmodern virtual art as Popper does, this chapter is focused on movement that is used to perceive the virtual as a means to produce material and immaterial effects, and to contribute to a consciousness of time that resonates with Moholy-Nagy, Gabo, and Bergson's understanding of the virtual. My approach cleaves new possibilities for considering a history of contemporary kinetic art history that does not move from mechanical kinesis towards immaterial digital art, as Popper suggests. Olafur Eliasson's *Your Negotiable Panorama* (2009) is one work that allows for such an approach to the virtual as it orchestrates movement in such a way as to build a consciousness of movement that is simultaneously material and immaterial in relation to the virtual.

Eliasson's work is also valuable for considering another element of Terry Smith's approach to contemporary art that in part resensationalises aspects of modern art. If, according to Smith, contemporary art has a specific relationship with temporality that requires 'many kinds of time from its audiences and offers many in return' (2011, p. 196) Eliasson contributes to contemporaneity by being primarily concerned with measuring the present as it passes. Eliasson echoes Bergson's modern conceptualisation of time: the present is articulated as an antiphrasis; as it unfolds it progresses, while memory in the present also gathers together and preserves the past. This chapter is therefore specifically engaged with this relationship between movement, the virtual, and the actual, and considers a continuation of movement instigated by Moholy-Nagy and Gabo and continued by Eliasson. In doing so, not only does Eliasson's work continue to use movement and time to draw his audiences into be conscious of the unfolding nature of time but, the artist uses these elements to consider the virtual as a part of this consciousness of time.

This chapter will begin by addressing the key understandings of Popper's concept of virtual art, and identify how he excludes virtual art from actual movement in art. This differs from Chap. 4 in that there the focus was on Popper's de-emphasis of kineticism in virtual art builds a distinctive rupture in the interpretation of kinesis in contemporary art, and restricts the virtual solely to a simulation of the real. Popper's use of the term 'virtual' in *From Technological to Virtual Art* is narrowly applied to a select range of digital art that was made since the 1990s, and, as I argued in Chap. 4, facilitates a technological determinism that excludes the virtual from other forms of media.

Secondly, I will present *Your Negotiable Panorama* as a continuation of engaging with the virtual that resonates with Gabo and Moholy-Nagy's use of the term. In this artwork Eliasson uses movement to explore the limits of perception, where images of movement oscillate between appearances of materiality and immateriality. This oscillation produces a specific consciousness of temporality as a constant state of becoming that accumulates the past within it. Eliasson's interest in the time as a construction of perception is emphasised to facilitate movement as a vehicle that extends the threshold of the present.

This chapter largely draws from Bergson's understanding of the relationship between duration and the virtual, as well as Deleuze's emphasis on the virtual as a chaotic field that becomes actualised in the present. Bergson's and Deleuze's perspectives are useful resources for bringing kineticism and the virtual together under the same discourse, and understanding some of the affects of movement in contemporary installation. In light of Bergson and Deleuze, I argue in this chapter that perceiving movement in works such as *Your Negotiable Panorama* can act as a visual gauge for negotiating the perception of the actual from the virtual, and for contemplating the crystalline images that arise when the distinction between the actual and virtual are indiscernible.

7.1 Frank Popper: From Kinetic to Virtual Art

'Virtual' and 'virtuality' are terms that have persisted throughout the history of kinetic art, often to describe the optical effects of movement from either static representation or actual movement. From as early as László Moholy-Nagy and Naum Gabo's experiments with form, movement, and light to produce 'virtual volumes', optical illusions were used to give mechanical media an appearance of weightlessness and alternative forms. Artists after WW2 also experimented with virtual volumes such as Jean Tinguely's collaboration with Yves Klein in *Translation No. 1, pour un triangle* (1958), Tsai Wen-Ting *Double Diffraction* (1972), Len Lye's *Grass* (1965), and Gerhard von Graevenitz's *Kinetic Object* (1965) with new media at the time. For many of these artists virtual volumes were a way to explore movement as a means for enabling mechanical media to be material and also weightless in appearance and enabling their work to oscillate between these two states (Burnham 1968a, pp. 236–243). Using movement to create a virtual volume encouraged viewers to perceive the edges of perception that transforms from stasis to movement, and to create images of the past in the actual present. For these artists the virtual resides in the present, and is an optical effect of movement to produce a blurring of past and present movement. What is unique to artists using kinesis to form experiences of virtual volumes is that the effect is primarily used to create forms that oscillate between materiality and immateriality.

Contrary to this approach to the virtual, Popper (1968) in *Art and Electronic Media* and *From Technological to Virtual Art* approaches the virtual in art as a means for simulating reality. For Popper and Nechvatal virtual art is comprised of 'elements from all the arts made with the technical media developed at the end of the 1980s' (p. 67), which enables the simulation of reality by encouraging experiences of immersion, interaction, and poly-sensorial engagements with digital representation. Popper's pinpoints the beginning of virtual art in a time where artists 'mastered' technological media such as holography, computer graphics, wearable interfaces, and screen based technologies to produce simulations of reality (2007, pp. 1–3). Popper's chronological distinction excludes modern industrial mechanical media because, according to him, it lacks the potential to be used to present simulations of reality. For instance, the optical, kinetic, and participatory art that Popper draws upon in *Art, Action and Participation* and *Origins* and *Developments of Kinetic Art* are focused on the actual technological and relational movements that occur in artistic experiences. Here, the motions in virtual art are primarily focused on those that simulate a constructed time and space, and mimic corporeal senses with digital representation (pp 181–220). This is why Popper explains that virtual art has little to do with the experience and perception of actual movement and the virtual signifies 'that we were in the presence not only of reality itself but also the simulation of reality' (2007, p. 2).

Popper continues that '[f]rom an ontological point of view, contemporary virtual art represents a departure from technological art since it can be realized as many different actualities' (p. 2). This approach to virtual art is largely influenced by the writings by Edmond Couchot, who has brought an awareness of the virtual as something that is opposed to what is actual, but can be translated by technological art through simulation. Both Popper and Couchot echo a Deleuzian vocabulary for these concepts to highlight a tension between the virtual and the actual. In Popper's words 'the virtual as a power opposed to the actual, but whose function, technologically speaking, is a way of being via digital simulation that can lead toward a certain expression of the operator's subjectivity' (2007, p. 3). Despite this technologically neutral definition of the virtual, Popper is adamant that exploring and representing the virtual in art in this way is strictly only possible through advanced technological media that emerged in the final decades of the twentieth century. This is evidenced in *From Technological to Virtual Art*, where Popper draws an explicit transition away from the study of movement in technological media that he outlays in *Origins and Developments in Kinetic Art*. In his interview with Joseph Nechvatal, Popper makes this clear by stating that movement 'is no longer a prerequisite for my interest in works of art' because virtual art is considered as a departure from kineticism (p. 71). In this regard there is little cross over between the movement in kineticism and the responsive, interactive, and tactile movements in virtual art and virtual reality.

Even though Popper gives attention to Moholy-Nagy and Gabo's early avant-garde virtual volumes in their art in the 1920s, they are only briefly mentioned as precursors to virtual art in *Origins and Developments of Kinetic Art*. Again, kineticism is used here to draw binary distinctions between analogue and digital media: Popper associates kinetic sculpture with the mechanical and material, while digital art is considered as being media for virtual and immaterial art. Although Popper's argument services digital media art with a heritage of machinic art, it does so through a misinterpretation of how early avant-garde kinetic artists explored the concept of virtuality in their art. As I address in Chap. 5, Moholy-Nagy and Gabo create virtual volumes to enable their sculptural material to oscillate between tangible and optical forms, and highlight the embedded qualities of their mechanical works as holding simultaneous qualities of material and immateriality. The virtual is not considered as an additional dimension or simulation of reality, but embedded within the ontological realm.

Consequently to this technological and chronological distinction, Popper also understands virtual art as the most technically demanding and sophisticated form of technological art, and stands as the most refined means for humanising technology. Even though he and Joel Slayton claim that virtual art has been deliberately constructed to have open and ambiguous definitions (2007, p. ix), Popper's digital determinism greatly restricts discussions of the virtual being derived from mechanical kineticism. This is particularly significant because his scholarship on the field of kinetic art since 1968 has been the key framework in which kinetic art history has been interpreted. Despite this, Popper's technologically determinist view also provides an opportunity for considering alternative approaches to the virtual in relation to kinetic art in order to discuss the roles and effects of kinetic sculpture in contemporary art.

7.1.1 Your Negotiable Panorama

Installed at the ARKEN Museum on the outskirts of Copenhagen *Your Negotiable Panorama* is a work like many of Eliasson's: a machine of reflection and refraction that the artist uses to explore alternative perceptions of movement, space, and time. When approaching the installation viewers are invited to walk over a ramp installed at the entrance of a circular space that has been sectioned off from the rest of the open gallery. This action triggers a pump that sets a pool of water at the centre of the space in motion. Installed above the water is the source of light projection: A glass prism and an HMI lamp, which illuminates, reflects and projects the ripples of the water onto the wall, producing a kinetic wave horizon that moves rapidly, before gradually calming down and creating smoother ripples along the walls. When viewers step onto and over the ramp they activate the movement and projection of water and, as a result, the form of the horizon on the walls articulates the weight, duration, and dynamic of the viewer's step. As Eliasson describes, it is their 'negotiable panorama'.

Art historians such as Eve Blau have previously drawn parallels between the works by Eliasson and Moholy-Nagy. In ways not dissimilar to Moholy-Nagy's *Light Space Modulator*, Eliasson brings attention to the space that surrounds the viewer by using projection of light and movement, and de-emphasising the contraptions that create and project the light forms. In both cases kineticism is the primary focus of the experience, rather than the object of movement (which in Eliasson's work is the moving water). Like Gabo, Moholy-Nagy, and Bergson before him, Eliasson uses space as a coordinate of movement and compels his audience to focus on their perception on the movement of the horizon and to monitor its flow and changing rhythms before it settles. Each person's stride, weight, speed, and/or hesitation when walking over the ramp effects how the light wave moves, as well as its duration for movement before returning to stillness. The movement that they perceive in the present has a dual kineticism, as it is an unfolding mediation of the reflection and refraction of light, as well as a response to the viewer's passage over the ramp that activated its movement.

Like Moholy-Nagy, Eliasson regards light as a carrier of motion; rather than using light to illuminate or signify forms, it is used to bring attention to the dynamics of perceiving motion. However, unlike Moholy-Nagy and Gabo, Eliasson's light is not used to study the effects of light that give the impression of a weightless, mechanical sculpture in movement. Instead Eliasson focuses on the materiality of light itself. This has been previously suggested by Mieke Bal who stresses that Eliasson often uses light as its own medium. In Bal's words: '[l]ight is not the opposite of materiality. Rather, Eliasson's rich exploitation of light achieves a reconciliation of matter and light: he invites us to experience the very materiality of light itself. Light, if taken seriously is a material element and not just as a tool for visibility' (p. 169). Bal's interpretation of Eliasson's use of light and movement conflicts with earlier interpretations of kinetic lumia that emphasise light as an inherently ephemeral and immaterial medium. Instead, Eliasson's light occupies both material and immaterial effects.

Like many of his artworks constructed with water, light, and motion, Eliasson imitates and abstracts the perceptions of natural phenomena in *Your Negotiable Panorama*. Other works by the artist involve reconstructions of waterfalls, rainbows produced by artificial, indoor rainfalls, the manipulation of perception through prisms, horizons, and reconstructed suns that loom over cityscapes or museum halls. In the case of *Your Negotiable Panorama* viewers are required to renegotiate Euclidian rationalisations of form and perspective with an artificial and moving horizon. Eliasson orchestrates these experiences to highlight the perceptual idiosyncrasies that occur in natural and urban environments to encourage viewers to perceive them in new ways. By isolating the experience of a horizon and resituating it in a gallery setting, Eliasson requires his viewers to relinquish the visual baggage that is normally associated with these everyday experiences, and reexamine their movements and manipulation of sensation and perception in a new environment.

Mirjam Schaub has argued that Eliasson experiments with the empathetic relations between viewer and object without attempting to humanise technology. Rather than making 'the physiological more "human"'. Schaub argues that Eliasson tends to asks viewers, 'to think of it from the perspective "of the objects", the "things" and their 'colors' that it produces *as if they actually existed outside of one's own retina*: an art of after images, trick images, of the reversal and immobilization of movement'(2009, p. 2). For Schaub, Eliasson's kinesis presents a spectacle of perception, rather than of mimesis.

Additionally, Mieke Bal considers Eliasson's intrigue with movement has a direct consequence on the individual perception of time. She suggests that Eliasson makes his viewers engage with the perception of time anew by 'shifting time away from its 'natural' space so that our belatedness becomes the primary condition of existence' (*Take Your Time* p. 173). Abstracting the movement of natural phenomena and centralising the viewer's perception of them in the gallery space erodes the conventional mechanised rationalisations of time. As Eliasson describes:

> [t]aking one's time means to engage actively in a spatial and temporal situation, either within the museum or in the outside world. It requires attention to the changeability of our surroundings. You could say that it heightens awareness of the fact that our actions have a specific speed, depending on the situation (in Bal and Grynsztejn 2007, p. 52).

If, as Terry Smith suggests, contemporary artists often require many kinds of time from their audiences in order to offer many in return (2009, p. 195), Eliasson offers a present that is centred on the subjective awareness of time at the centre of the experience. In exchange for giving one's time when viewing the installation, Eliasson abstracts, reengages and directs the viewer's attention onto the present to heighten sensations and perceptions of time as a state of constant transformation. The artist also continues a tradition within contemporary kinetic art history by using movement to draw attention in towards time as a constant transition that is changeable in its rhythms. Eliasson encourages a phenomenological perception of time that implicates the subjective perceptions of the viewer at the centre of the installation.

As described by Pamela Lee, 'Eliasson has produced a body of work that variously engages questions of subject-object relations, exploring the ways in which the subject's encounter with his or her surroundings prompts larger revelations about the nature of perception itself' (Lee 2007 p. 34). In the case of *Your Negotiable Panorama* the installation that surrounds the viewer is a mediated mirror of the viewer's movement in order to bring the viewer's durational experience into the centre of the work as a site of privilege, possession, and transformation (Blom p. 20). For instance, the work and titles of *Your Only Real Thing is Time* (2001), *Your Strange Certainty Kept Still* (1996), *Your Waste of Time* (2006), and *Your Wave Is* (2006), *Take Your Time,* (2008) and exhibitions such as *Your Intuitive Surroundings Versus Your Surrounded Intuition* (2000), indicate that transformation occurs as the work unfolds-with the viewer's subjective experience.

Eliasson pursues a phenomenological approach to his work in order to emphasise the experience of 'perceiving yourself perceiving' space as a coordinate of movement. In doing so Eliasson's experimentation is useful for considering further the implications of movement as an incipient and emergent process, as well as considering art historical understandings of the virtual that have been used to build a consciousness of time through movement. This is because Eliasson's understanding of the virtual is geared according to the process of becoming actualised in the present.

This perspective is also affirmed by Eliasson who described his interest in phenomenology as an ongoing theatrical dynamic between the object and the viewer's perception of it. He says, '…I don't think that the object actually *does* anything. But it is also not that I am saying that the subject is everything. I think that the subject has an impact on the object, and vice versa, the object has an impact on the subject' (in Morgan 2011 p. 17). As an example of this intention viewers of *Your Negotiable Panorama* are surrounded by the wave horizon. This wave is a translation of their own movements over the ramp in order to enable '[their] experience of the thing [to be] integrated as a part of the thing itself' and to find a bleed between the subject's actions and their perceptions of the installation (Bal and Grynsztejn 2007, p. 54).

Smith has referred to Eliasson as an artist who resides in his third current of contemporary art—one that 'manifests the conditions of contemporaneity—who rejects grand statements and finds himself exploring ways of taking small, but hopefully significant, steps within this seemingly limitless stream of times' (2009, p. 198). However, Eliasson also resensationalises modern approaches to art, particularly through his emphasis on presenting a consciousness of time through his work. As with modern artists before him Eliasson is focused on the ceaseless shifting of time from the past into the present. Eliasson emphasises this continuity of time in *Your Negotiable Panorama* by projecting the movement rhythms of the water out onto the surrounding walls. The wave foregrounds movement and time as continuous and changing in its rhythm. It coincides with the expansion and contraction of perception, action, and sensation, and pulls the viewer's attention towards the present as a construct of perception. The present is an important feature of *Your Negotiable Panorama*, as well as many of the artist's other works. As Eliasson describes through an analogy, he aims to draw together his installations as events of the present temporality:

> There is only a 'now'. But I think that people might believe that there is a time. Let me try to explain through an example: if I am sitting in a boat, like I did this summer, going down a river, I am 'now' in the boat, at this spot on the river, and the landscape on the banks passes me as time. If I stand on the bank and the river passes me, the water which is further up the river is also 'now' even though I know that it is not yet here. Our belief in time is just a construct (Bal and Grynsztejn 2007, p. 16).

Because time is a construct of perception, or, perhaps more precisely, of imagination, the processes of perception and sensation, are active, rather than passive participants in the flow of time as it unfolds with them. In *Your Negotiable Panorama* the viewer's perceptions and actions are at the frontier of this now. Eliasson encourages his audiences to consider each action as a motion that pushes the horizon of the present into a new field of actions. By making the projected horizon on the wall responsive to the viewer's motion, its movement in the present is continuous and relative in nature as each step and wave of the horizon brings attention to the action that activates space and time and affects its surrounding variables. Each rhythm in the wave corresponds to the dynamics of each individual step on the ramp to highlight that action produces a ripple effect of responding motion.

Considering Eliasson's position on kineticism, time, and perception, a Bergson-Deleuzian interpretation of *Your Negotiable Panorama* is useful for understanding the incipient nature of movement that is in a process of actualisation from the virtual. Bergson and Deleuze's understanding of time, memory, and the virtual are also useful for considering kineticism in art as an ongoing contemporary practice that diverges away from a digitally determinist reading of the virtual.

Additionally, this analysis also demonstrates that kinetic dynamism is a useful tool for discussing the facets of Bergson and Deleuze's understanding of the relationship between the actual and the virtual.

7.2 Intersections with the Virtual: Eliasson, Deleuze and Bergson

As I highlighted earlier, Popper echoed Bergson's approach to the virtual and duration when analysing kinetic art. However, despite this, there is a distinctive difference between Deleuze and Bergson's approaches and Popper's. While Popper considered the virtual as a simulation that exists in the actual, ontological realm, Deleuze drew from Bergson to regard the virtual differently to the actual, that becomes actualised in the present. In *Bergsonism* the relationship between the virtual and actual is described as:

> The possible has no reality (although it may have an actuality); conversely, the virtual is not actual, but *as such, possess a reality*.…The virtual on the other hand does not have to be realized, but rather actualized; and the rules of actualization are not those of resemblance and limitation, but those of difference or divergence and of creation (pp. 96–97).

Deleuze (1991) makes it clear that while the virtual is opposed to the actual, the virtual also 'possesses a full reality by itself' and this reality is not a unity but 'designates a pure multiplicity' that is yet to be actualised in the present (*Difference and Repetition* 211). Because of this, the relationship between the virtual and actual is in a continual process of actualisation, or becoming. Simon O'Sullivan has interpreted this relationship between the virtual and the actual clearly by emphasising the virtual as a state of becoming:

> The possible names a logic of Being (ontology of stasis), the virtual affirms a logic of becoming (ontology or process). […]The virtual, or rather the actualisation of the virtual, is then the creative act—precisely the production, or actualisation, of difference and thus diversity from a pre-existing field of potentialities (2006, p. 103).

This field is a chaotic realm that becomes unified in the present when actualised. The process of actualisation is also a continual process and, as Bergson described, 'the virtual image evolves towards the virtual sensation, and the virtual sensation towards real movement' (1911, p. 168). It is also important to note that this transition from the virtual to the actual occurs by degree. For Bergson realising this transformation from the virtual to the actual also brings attention to the sensations that have 'tried to embody itself in the sensation' (p. 169). In this regard an affective potential of kinetic movement lies in its ability as a medium to elaborate on the incipient nature of the actual from the virtual. That is, kineticism can be used to emphasise the temporal aspects of the virtual because it brings attention to movement as incipient and emergent, and gives rise to the perceptual edge between the natural continuation of the present, and what virtual states are yet to be embodied in the actual.

When interpreting *Your Negotiable Panorama* from Deleuze and Bergson's understanding of the virtual, rather than a more strict phenomenological approach, the installation acts as an event horizon of actualisation that brings attention to the relationship between the virtual and the actual. Eliasson's horizon of light and water articulates force through the actions of the viewers as well as the systematic coordinates of light, water, movement, prism, and projection that come together to produce the horizon on the wall. Motion highlights the process of actualisation as continual rather than discrete, and is a process of becoming that is uncertain, unstable, and yet constant. The movement of the projected horizon produces a sense of the actual being drawn from what O'Sullivan has highlighted as the 'pre-existing field of potentialities' (p. 103), because of its unfolding durational rhythm. In this regard, Eliasson encourages his viewers to move with what Brian Massumi describes as a manner of seeking 'the seeping edge of perception' as the horizon moves (2002,p. 217).

The horizon is not only a presentation of movement but, also a transduction of energy that has moved from the viewer's actions onto the ramp, and to the centrepiece of the installation. The refraction and projection of the wave unfolds in the present, yet it is also an echo from the past. For the artist, this is much like how memory is activated and recalled in the present: '[t]o say that memories are just representational does not do justice to the memory. Sure, memories are representational in the way that they are stored, but a memory only exists in the now [...] You can experience something in the past. So memory is representational but it is also now—it has this ambiguity' (Morgan p. 22). Eliasson points out the antonymous nature of memory that can be applied to the wave in *Your Negotiable Panorama*, where it is a wave that is unfolding in the present but, at the same time, a stored continuity from the past. Considering Eliasson's description, the present is rich with folds of the temporalities within it, as it also unfolds.

This orchestration of movement resonates with Bergson's concept of duration as a continual, unfolding process of creation and differentiation. For Bergson, 'duration means invention, the creation of forms, the continual elaboration of the absolutely new' (1913, p. 11), which Eliasson explored with the panoramic horizon. By positioning the subject's perception at the centre of the installation, and their actions at the frontier of the experience of the installation, Eliasson highlights experiential time as a creative process where action propels the actual and pulls the virtual into the present.

Another way to consider the virtual in relation to the actual is through Bergson's elaboration of memory. In his writings, Bergson argues that the past is coextensive with the present (1911, pp. 105–109). The virtual is not only what is yet to be actualised, but also what has passed from the actual into the past, in a process of deactualisation. Memory, then, is a way of recalling the past and bringing it into the present. In this sense, 'memory does not consist in a regression from the present to the past, but, on the contrary, in a progression from the past to the present' (1911, p. 319). As O'Sullivan has highlighted, the 'virtual can be understood as a temporal dimension of the object' (p. 209), because the actual is not only an unfolding of the present, but is also a site for preserving the past within it.

Eliasson echoes a similar attitude when he explains that perception in the present is informed by the accumulation of sensations that are built by memory:

> [t]he sense of time that I work with is the idea of a 'now'. I would say that there is a timeline, which is obviously divided by 'now' and the past and the future. But I don't think it is really possible to talk about the past and the future—however, maybe it is possible to talk about memory and expectations. My 'now', my sensation of now, comes from the idea of the subject from which it derives. I can say that my past and future are 'now', for the world (in my work). Or *your* memory and *your* expectations are now for me (Morgan p. 16).

To experience *Your Negotiable Panorama* is to engage with a consciousness of temporalities that explicitly fold together, rather than being experienced on a linear plane. Eliasson presents both a stored continuity of time as well as a performance in the present in the installation. This is achieved by making the horizon perform movement in the actual present, while also translating a past action. The artist explains that the slippage between the past and present in his works articulates the antonymous nature of the present because it is both unfolding as it progresses, as well as recalling the past within it. Eliasson has attributed a direct interest in Bergson's understanding of the relationship between the present and the past embedded within it, and explained that '[t]o me the richness of Bergson's thought lies in his idea of temporal density. He doesn't see time as a line; time is non-linear, and he would say that space is *of* time, not *in* time. One can't talk about matter at all if one doesn't take the temporal aspect into consideration' (Engberg-Pedersen 2008, p. 301). Eliasson's movement of the horizon is an echo from the past, unfolding in the present and explores this temporal density of Bergson's duration.

As Bergson argued, the problem with representation is the inability to store continuity. Bergson also emphasised that the representation of movement is only produced due to the actual movement of the projector (1913, p. 322). Actual kinesis in sculpture and installation, however, performs this continuity, thereby avoiding the polemic of distorting duration. However, the kinetic dynamism utilised by Eliasson brings this polemic between Bergson's duration and representation to the fore in *Your Negotiable Panorama.* The kinesis in this installation performs a moment of movement that highlights the emerging and yet ungraspable nature of the present. The wave of the horizon is a translation of the viewer's passing over the ramp, as well as performing a transduction of energy, from the viewer to the projected walls. While Eliasson has constructed this space specifically to react to the viewer's entrance in the room, he does so to demonstrate that all movement produces ripple counter effects of motion. While the wave makes this perceptible, it acts to allude to the relational movements between particulars, which inevitably occur in duration.

Eliasson's installations are not simulations of the natural world, but they are what Boris Groys would suggest of contemporary installation: a way of reterritorialising temporality in art. For Groys, contemporary installation art is increasingly created to help express the nature of contemporaneity in contemporary art (pp. 71–80). Groys (2008) argues that each reproduction or reconstruction is never entirely a repetition of the work, because each viewing of the installation is a new encounter that pulls new perceptions, associations, and reactions from the work.

For Erin Manning, to 'see yourself seeing' uniquely describes the perception of movement as it unfolds (2009, pp. 5–6). To feel movement as it forms, is an act of looking with form as it is moving, rather than looking-for evidence of movement passing. Manning borrows from Bergson to emphasise that movement can never be accurately represented by media because each representation is an act of isolation and abstraction away from movement's rhythm as it comes into and out of form. This is why Manning argues that to feel movement moving, or in Eliasson's terms, to see yourself seeing movement, is to sense movement in the actual that emerges from an incipient state.

In this regard installation art has a unique potential to act as an assemblage that builds an experience more than an object. Eliasson uses this affective potential to encourage his viewers to 'see yourself seeing' in the present. The artist explains:

> What is special in the case of conditional experience is, I think, what I sometimes call the introspective quality of seeing: you see whatever you're looking at, but you also see the way you're seeing. You can find pleasure or fear in what you're experiencing, but your experience of the thing is integrated as a part of the thing itself. (Bal and Grynsztejn 2007, p. 54).

Eliasson's phenomenological description of the affects positions the subject at the centre of the experience; they are both the subject and object of the moment constructed. The movement of *Your Negotiable Panorama* then begins with the viewer at the centre of motion and action.

7.3 Deleuze's Crystalline Experiences of Temporality

Art historian Philip Ursprung has argued that Eliasson creates works in reference to a system of surfaces that are refracted and reflected when viewers encounter the installations. Ursprung proposes, 'I would call this structure "crystalline". It reveals itself through a specific, finely faceted surface that produces countless mirror images of anyone who moves in front of or inside it' (2008, p. 14). Some of Eliasson's works are indeed crystalline in appearance. They are installations that take the form of refracted mirrors, kaleidoscopes, and mosaics, while in other works Eliasson centralises perception to produce infinite images that enable the viewer to 'see themselves seeing' (Bal and Grynsztejn 2007, p. 55). As Ursprung explains, these latter works refract their surfaces, which 'could be described simply as the product of never-ending faceting, complex contortions and undulations, in which the image of the surroundings is broken down, prism-like, into an infinite number of new images in which every observer can ultimately recognise him- or herself' (p. 14). Drawing from Frederic Jameson, Ursprung considers the postmodern distribution of surfaces that constitute a hyperspace to be central to understanding Eliasson's work.

However, Ursprung's focus on the material and perceptual surfaces of Eliasson's artworks does not tend to the temporal layering that is constructed in *Your Negotiable Panorama.* While the installation works within Ursprung's rubric by presenting a mechanism that refracts images and creates a wave horizon, this interpretation is made with a de-emphasis of the dense folds between perceptions of the past, present, and memory in this installation. When considering the work from a temporal point of view, there is a stronger potential to discuss a Deleuzian approach to the crystalline. Doing so is useful for understanding further the relationship between movement in the actual in relation to the virtual, while also bringing forth new ways to discuss and interpret Deleuze's framework for the crystalline image. Additionally, the temporal layering that can be perceived in the experience of *Your Negotiable Panorama* is heightened further by using the crystalline as a tool for interpreting and discussing the movement of the installation.

While Deleuze discussed the crystalline regime through his interpretation of modern film in *Cinema 2: The Time Image*, his analysis contains key philosophical tools useful for the practical philosophy related to time-based media, events, and experiences. In contrast to Ursprung, Deleuze's crystalline image is primarily regarded through a temporal, rather than a spatial regime that is qualitative, non-linear, and folding, and produced purely through the perception of optical and aural stimulus. As I addressed earlier, Deleuze understood the virtual and the actual to be in opposition to one another, and yet involved in a continual process of actualisation from the virtual. There are, however, moments when the distinctions between the virtual and the actual coalesce with one another and become unclear. As Deleuze explained, '[t]he two modes of existence are … combined in a circuit where the real and the imaginary, the actual and the virtual, chase after each other, exchange their roles and become indiscernible' (p. 127). This indiscernible quality is a core aspect of the crystalline image. For Deleuze this new image does not collapse the seemingly opposing qualities of the virtual and actual, but, rather, can be used to describe when the differences between the actual and the virtual cannot be located.

The crystalline is more than an unclear perception of the distinctions between the actual and the virtual, but a separate regime of images. Deleuze explicitly acknowledges this when he describes that the crystalline 'is completely different' from the actual and virtual. He continues, 'the actual is cut off from its motor linkages, or the real from its legal connections, and the virtual, for its part, detaches itself from its actualizations, starts to be valid for itself'(1989, p. 127). The crystalline therefore resides at the intersection between two systems of images, the organic (containing the virtual and the actual) and crystalline images. Perception can be mobilised to enable 'passages from one regime to the other, from the organic to the crystalline, can take place imperceptibly or there can be constant overlapping' (p. 127), and challenge a Euclidian system of representation.

When considering Deleuze's use of the crystalline, it becomes apparent that Eliasson takes advantage of a system of perception that folds, coalesces, and blurs images of temporality in *Your Negotiable Panorama.* The projected panorama forms and moves in the present, but it does so as a signification of the viewer's step off the ramp that occurred previously. Although it is clear that the movement of the water is occurring in the same time and space as the viewer (and because of this the installation falls short of a purely Deleuzian crystalline image), Eliasson uses the movement of the water to foreground the temporal folds that exist within the present. If the wave can be perceived as a representation of the past and simultaneously also a presentation of the present, there is an intersection between a perception of the past and, also, a perception of the present within the same image of movement. Furthermore, because of Eliasson's experimentation with Bergson's approach to the present as a process of unfolding time, and preserving the past within it, Eliasson highlights the multiplicity of temporality and the folds within it.

7.4 Conclusion: Resensationalising Kinetic Trends

My approach to *Your Negotiable Panorama* is a diversion from Popper's understanding of the virtual and virtual art, where he and others focus on immateriality and a simulation of reality. Emphasising the temporal aspects of the virtual is consistent with Moholy-Nagy and Gabo's experiments with their virtual volumes and suggests multiple approaches to the virtual in kinetic art. In *Your Negotiable Panorama* Eliasson is careful to emphasise the perception of movement as it is actualised in the present in order to highlight the virtual as a part of the real, rather than a simulation of it. Therefore, while Popper suggests that early avant-garde kinetic art is a 'starting point for the context of high technology art'(1993, p. 65), such as what he regards as virtual art, my analysis of Eliasson's installation in reference to my early approach to Moholy-Nagy and Gabo cleaves a new connection between avant-garde and contemporary kineticism with the virtual. Therefore, as I have suggested, artists working with kineticism have the potential to examine the relationship between the virtual and the actual in time that moves with the viewer. What is unique to Eliasson's *Your Negotiable Panorama* is his ability to use movement, projection, and representation to also express the present as an antiphrasis that folds the past within it while it also unfolds through a crystalline regime of images.

Eliasson therefore contributes to the *peculiar time* of contemporary art and society by resensationalising Bergson's modern concept of duration in a contemporary technological context. This not only contemporises Bergson's approach to time, movement, and the virtual but also illuminates an historical pattern in kinetic sculpture and installation that uses movement to make machinic media oscillate between materiality, immateriality, the virtual, and the actual. The motion rhythms that emerge in the experience of *Your Negotiable Panorama* therefore highlight motion as an effective tool for exploring the nature of perception. This chapter has focused on how Eliasson uses movement to require a self-reflexive renegotiation of perception, time, and space, and can be used to encourage the material and immaterial properties of motion in both digital and analogue media. Eliasson renegotiates movement in relation to the virtual that situates perceiving yourself perceiving movement as it arises from the virtual and forms in the actual present.

This present is antiphrastic in nature; as it is a visualisation of the present unraveling in time, while also preserving perceptions in the past within it.

Additionally, discussing movement in the actual as an incipient process that emerges from the chaotic virtual plane is useful for contemplating Deleuze's crystalline regime of images, as well as Bergson's argument that the memory in the present preserves the past within it, as it unfolds. While Eliasson is explicitly interested in constructing phenomenological experimentations of perception, his works such as *Your Negotiable Panorama* also visually resonate with Bergson and Deleuze's approach to the virtual. As an image of time, Eliasson encourages a perception of the crystalline that intersect on a temporal, rather than a spatial plane. This is useful for considering relationships between kineticism and the virtual that diverge from nurturing virtuality as a visual effective of digital representation and simulation of reality as suggested by Popper in *From Technological to Virtual Art* and *From Technological to Virtual Art.*

This temporal reading of Eliasson's work is an investigation of the implications behind Smith's description that there is a tendency in contemporary art for artists to recall and perpetuate modern tropes. In the case of Eliasson I have addressed *Your Negotiable Panorama* in order to elaborate on contemporaneity through a Deleuze-Bergsonain understanding of temporality as a constant becoming that preserves the past within it. Crystalline images are used to create indiscernibility between images of the past and present as they emerge from a state of incipience. Eliasson's installation therefore also contemporises Bergon's modern conception of duration. Smith's remodernism is also useful for reconsidering the dominant tendencies within contemporary art history of kinetic sculpture, particularly Popper's regard for kinesis in regard to the virtual. By considering the material and immaterial affects of the virtual, kinetic sculpture is not only considered as a mechanical precursor to digital art, as Popper suggests, but also an avenue for exploring the changing conceptions of

References

Bal E, Grynsztejn M (eds) (2007) Take your time: Olafur Eliasson. Thames and Hudson, San Francisco Museum of Art

Bal M (2007) Light Matter. In: Bal E, Grynsztejn M (eds) Take your time: Olafur Eliasson. Thames and Hudson, San Francisco Museum of Art

Bergson H (1913) Creative evolution, (trans. Mitchell A.). Macmillan, London

Bergson H (1911) Matter and memory, (trans. Paul N. M. and Palmer W. S.). Macmillan, New York

Burnham J (1968) Beyond modern sculpture: the effects of science and technology on the sculpture of this century. G. Braziller, New York

Deleuze G (1991) Bergsonism, (trans. Tomlinson H. and Habberjam B.). Zone Books, New York

Deleuze G (1989) Cinema 2: The time-image, (trans. Tomlinson H, and Galeta R.). University of Minnesota Press, Minneapolis

Engberg-Pedersen A (2008) Studio Olafur Eliasson: an Encyclopedia. Taschen, London

Groys B (2008) The topology of contemporary art. Antinomies of art and culture; modernity, postmodernity, contemporaneity. Duke University Press, Durnham

Lee P (2007) Your light and space. In: Grynsztein M. (ed.) Take your time: Olafur Eliasson. Thames and Hudson, San Francisco Museum of Modern Art

Manning E (2009) Relationscapes, movement, art, philosophy. MIT Press, Cambridge, Massachusetts

Massumi B (2002) Parables for the virtual: movement, affect, sensation. Duke University Press, London

Morgan J (2011) Interview with Olafur Eliasson. In: Morgan J, Lerup L, Morgan J (eds) Olafur Eliasson: your only real thing is time. Institute of Contemporary Art, Boston

Nechvatal J, Popper F (2004) Origins of virtualism: an interview with frank Popper. Art J 63 (1):62–77

O'Sullivan S (2006) Art encounters Deleuze and Guattari: thought beyond representation. Palgrave MacMillan, New York

Popper F (2007) From technological to virtual art. MIT Press, Cambridge, Massachusetts

Popper F (1968) Origins and developments of kinetic art. New York Graphic Society, New York

Popper F (1993) The place of high-technology art in the contemporary art Scene. Leonardo 26 (1):65–69

Schaub M (2009) The logic of light: technology and the humean turn. In: Eva Ebersbeger, Zyman D (eds.) Thyssen-Bornemisza art contemporary: the collection book Walther Konig Verlag, Cologne, pp. 138–151

Shanken E (2009) Art and electronic media. Phaidon Press, New York

Smith T (2011) Contemporary art: world currents. Prentice Hall, Upper Saddle River

Smith T, Enwezor O, Condee N (eds) (2008) Antinomies of art and culture; modernity, postmodernity, contemporaneity. Duke University Press, Durnham

Ursprung P (2008) Olafur Eliasson's studio. Taschen, Cologne

Chapter 8
Conclusion: Consciousness of Time: Looking Back and Moving Forward

> *Art today is shaped most profoundly by its situation within contemporaneity....*
> *Contemporaneity itself has many histories, and histories within the histories of art.*
>
> (Smith 2009, p. 6)

This book has focused on how actual movement in art has been used to influence a consciousness of time in contemporary society. The practice of using movement to orchestrate perceptions of time has been a longstanding feature of contemporary art history. For instance, from as early as 1919 Gabo encouraged his audiences to seize the present in his *Realistic Manifesto* and created *Kinetic Construction: Standing Wave* as a sculptural experiment that uses movement to bring attention to and create new perceptions of time. Additionally, Tinguely used his auto destructive sculptures to emphasise time as something that is inherently unpredictable, and which is rationalised and regulated by technology, while Olafur Eliasson focused on the subjective perceptions of 'now' in the kinetic rhythms of *Your Negotiable Panorama*. Throughout this book I have drawn from the works and writings of Gabo and Tinguely, as well as McCall, Kapoor, Haacke, and Moholy-Nagy, specifically because each of these artists use movement to orchestrate being with time in different ways. This book has therefore been broadly concerned with the roles and effects of kinetic sculpture as a means for engaging with the changing conceptions and expressions of contemporaneity. This tendency to use movement to express time, as I have argued, contributes to a wider consciousness of different technological ages each rationalising a specific peculiar time.

Chapter 1 addressed the view that avant-garde kinetic sculpture and installation has recently experienced a renaissance in contemporary art institutions. A number of large-scale exhibitions have returned to avant-garde kinetic artists, and, at times, exhibited their works alongside contemporary artworks in order to reflect on the continued, and yet changing expressions of movement in real-time. There have also been a number of contemporary exhibitions that have brought attention to earlier artistic experiments with kinesis in order to seek new relationships between a history of kinetic art in relation to the contemporary. This recent trend suggests that

© Springer Nature Singapore Pte Ltd. 2017

C. Chau, *Movement, Time, Technology, and Art*, Springer Series on Cultural Computing, DOI 10.1007/978-981-10-4705-3_8

there is attention to avant-garde kineticism in the wider milieu of contemporary art for a variety of reasons that contribute to a wider understanding of contemporary art history and facets of what it means to be with time.

It is the contemporary art historical perspectives of kinetic art that this book is most focused on. A contemporary interpretation of kinetic art is useful not only for considering the changing expressions of time in art, but, also, for considering how these artworks are framed and interpreted by contemporary perspectives, which differ from those previously held. In doing so, this book contributes to a contemporary consciousness of time in art. The term 'kinetic' has therefore been used in a self-reflexive manner in order to examine various ways in which kinetic sculpture and installation have, at times, been pursued and marginalised by art theory, criticism, and art history. Perhaps rather than considering that kinetic artists failed and became unrequited in their aims, as Burnham has argued, which Chap. 3 focused on, kinetic sculpture and installation has succeeded as a contemporary and historical experimentation with technological media to form expressions of time with movement.

I drew upon Sydney's Sixteenth Biennale titled, *Revolutions—Forms that Turn* as one example of how contemporary art is, in part, characterised by drawing on the recent past to unpack and reflect on contemporary identity. I argued that this exhibition drew upon kinetic art to reflect on the changing expressions of temporality and contemporaneity through multiple technological ages. I highlighted that the exhibition drew from avant-garde and contemporary artists to examine the term 'contemporary' as a continual accumulation and preservation of art history in the present. Contemporary art history is therefore an amassing of previous historical modes, rather than a resistance or departure from modern and postmodern frameworks. While the biennale exhibited a range of contemporary and historical artists associated with a variety of modern movements such as Futurism, Constructivism, Minimalism, and Dadism, my focus on the effects of incorporating kinetic artists within the exhibition is used to engage with expressions of time through movement in contemporary art history. My emphasis on the kinetic artworks in the biennale is valuable for considering kineticism to be at once, contemporary and historical. As noted at the start of this book, Christov-Bakargiev takes the view that: 'everything that exists in the world is of my time' (p. 33). Christov-Bakargiev therefore curated historical avant-garde works alongside contemporary pieces, in order to communicate that what is recalled by contemporary art history is also contemporary. In the scope of *Revolutions- Forms that Turn,* the works included in the biennale therefore inform audiences to reflect on what it means to be contemporary today.

Revolutions—Forms that Turn helps to raise questions around how and why avant-garde kinetic art is often interpreted as an historical, antiquated, and obsolete practice, particularly when movement continues to be used in sculpture and installation by contemporary artists. Rather than regarding avant-garde kinetic art as a practice that is forgotten, early mechanical experimentations of movement were curated in the Biennale to gather in avant-garde understandings of movement, time, and the machine as a continued concern related to contemporary identity. My approach to *Revolutions—Forms that Turn* also highlights a wider need for further discussion around the roles and effects of kinetic art and its relationship to temporality. For instance, if avant-garde and contemporary kinetic artists contribute to an understanding of contemporaneity today, what are the reasons that kineticism has generated claims of no longer being with time and being an obsolete, unrequited, forgotten and antiquated practice. Furthermore, exhibitions such as *Revolutions Forms that Turn* bring into question how avant-garde kinetic artists contribute to a consciousness of time today.

Chapter 2 draws on the arguments presented by art historians Lee and Smith who claim that the contemporary relationships to time are influenced by the turbulent social, technological, and political changes in Europe and North America during the 1960s. For Lee, '[t]he sixties are endless. We still live within them….The Sixties are endless in staging endlessness as a cultural phenomenon…This is one legacy of sixties art that continues to haunt us today' (p. 259). Like Lee, Smith argues that not only is the reflection of past expressions of conceptions of time in art informant of contemporaneity today but, this is also further complicated because contemporaneity during the 1960s is marked by an evasive and contradictory nature. In light of Smith and Lee I argued that the 1960s is marked by a turbulent relationship to time. I draw upon Michael Fried's seminal essay that objects to the role of duration in art in his 'Art and Objecthood' as a prime example of the trepidation around time in art. I also drew upon the auto-destructive artworks by Tinguely, *Homage to New York* (1960) and *Study for an End of the World No. 2* (1962), to articulate the turbulent consciousness of time in art. While the reasons for this turbulence are addressed, I argued that kinetic artists such as Tinguely contributed to a turbulent consciousness of time by orchestrating new perceptions of time with mechanical and tele-communicational media. While both works are auto-destructive in nature, I focused on the differences between these two works to argue that Tinguely's kineticism is sensitive to the ways that different technologies can be used to rationalise time in different ways, and that, at times, can also be incompatible with one another. The combination of Tinguely's artworks, Peter Selz's curatorial direction in *Directions of Kinetic Sculpture,* and Fried's attack on the incorporation of duration in art describes a period of temporal turbulence that has extended through to contemporary perspectives of time and art.

Another key aspect of the approaches to kinetic art history today are the influential writings on the intersections between art, science, and technology by Burnham. In Chap. 3 I addressed Burnham's 'Systems Esthetics', 'Real-Time Systems', and *Beyond Modern Sculpture* as central texts that have influenced the historical approach to kinetic art in relation to contemporaneity today. As Jones has recently explained in *Artforum*, Burnham's argument in 'Systems Esthetics' has come to embody many key characteristics of contemporary conceptual art today (113–114), and is an influential text for contemporary media art history. Across these texts Burnham describes kinesis in art at the time as an 'unrequited' machinic practice that failed to be relevant to the emerging postmodern approaches to interpreting and producing art, such as the appropriation of general systems theory. For Burnham, Ludwig von Bertalanffy's *General Systems Theory* was a key influential concept that depicted key aspects of society during the 1950s and 1960s.

Burnham's appropriation of systems theory was used to propose a new way of producing, engaging with, and critiquing art that was emerging at a time that, Burnham argued, kinetic sculpture was no longer relevant.

Burnham's criticism of kinesis in art strengthens an association between kinetic sculpture and modern antiquity because he associates kineticism with a machine aesthetic rather than an exploration of movement across a variety of media. Instead of affirming an incompatibility between kinetic artworks and systems theory, I demonstrated that the works by artists such as Haacke, can be understood as works that express temporality systems and systematic processes of perception. Haacke's early systems art depends on movement to 'physically communicate to one another' within the gallery space to highlight immaterial and material effects of perceiving movement in art in ways similar to Burnham's system aesthetics.

Haacke's use of actual movement resists traditional understandings of a work of art as autonomous, finite, and a product of craft by performing movement in real-time with the viewer. By identifying Burnham's approach to kineticism as a polemic in contemporary art history, Chap. 3 focused on the subtle, yet important connections, between systems theory and the study of movement in sculpture and installation. This analysis presents the use of form and movement as tools for expressing temporal systems and systems of perception by the viewer in real-time. My analysis of Haacke's artwork and approach to Burnham's argument in 'Systems Esthetics' is valuable for pursuing new approaches to movement and time from a systems perspective. My interpretation of Haacke's art emphasises that movement and time are key aspects of the artist's approach to systems theory, and is useful for making viewers conscious of the systems of perception at play when confronted with ontologically unstable works of art.

Chapter 4 identified Frank Popper's scholarship on kinetic art as another polemic within contemporary art history. Here I highlighted that Popper's extensive understanding of avant-garde kinetic sculpture in relation to other artistic practices leaves little room for considering the role and function of kinetic sculpture in ways that move away from building a genealogy of digital media in art. In this chapter focused on Popper' arguments in *Art, Action and Participation* and *From Technological to Virtual Art* to bring attention to his approach to kinetic art as a technological, formal, and conceptual precursor to subsequent media and conceptual art practices.

As part of this I pointed to Popper's attention to demateriality in art as a significant polemic for contemporary kinetic art historical study because he associates movement in art with materiality, the machine aesthetic, and formalism. Popper's argument locates kinetic sculpture as a modern mechanical formal performance of movement that is separated from conceptual and participatory artworks that incorporate movement. This is because, according to Popper, artists in the 1970s became increasingly interested in the conceptual properties, and disregarded the ontological elements of their art. In doing so, communicative, relational, and digital movement became a primary concern in art, rather than experiments with the affect of ontological movement as a focal point. However, Popper considers conceptual art and demateriality to entirely dismiss the effects of form in art. However, while dematerality indentifies a separation between the formal and conceptual properties of art, art historians such as Lucy Lippard have drawn attention to the importance of form to act as a gateway for the concepts behind a work of art. Popper's approach also induces a technological determinism that aligns contemporary digital media art with immateriality and a focus on conceptual properties of a work of art.

In order to demonstrate that Popper created a digital determinism that prevents kineticism from being discussed according to its material and immaterial qualities, I considered artworks *Line Describing a Cone* by McCall, and *Shooting into the Corner* by Kapoor and emphasised movement as an emergent process that is formed in an incipient state of preacceleration. This approach to movement opens the potentiality to discuss the emergent materiality of movement in art. My approach to McCall and Kapoor's kinesis destabilises Popper's binary distinction, and opens for discussion of the material and immaterial effects of movement.

Additionally a critical consideration of McCall and Kapoor positions their kineticism as a means to perpetuate modern conceptions of temporality, in part because they self-reflexively engage with breaking from traditional modes of representation in order to gain the attention of their audiences in the present. In this sense, McCall and Kapoor resensationalise temporality as something that is a constant transformation and is unified and eternal. This approach to McCall and Kapoor's work is important for considering the emergent, material, and durational capacities of kinesis in art, and destabilises Popper's use of demateriality, while bringing a consciousness of time and its material and immaterial effects to the forefront of kineticism.

Exhibitions such as *Revolutions—Forms that Turn* position early European avant-garde artists as central figures in contemporary art history. Artists of this period, such as Moholy-Nagy and Gabo, have also featured prominently in contemporary art historical exhibitions and are treated as early pioneers of the modern avant-garde kineticism. In cases such as *Revolutions—Forms that Turn* a return to the historical avant-garde was not only presented as an historical venture but, also a means to discuss and understand some key influences of contemporaneity in current society. These artists are therefore perpetuated by contemporary art history as influential figures that experimented and discussed the subjective perception of movement from which the ideas of temporality, particularly being with time, in contemporary society.

Chapter 5 of this book focused on three influential figures of this period who pioneered early experiments of sculptural kineticism: *Light Space Modulator* by Moholy-Nagy and *Kinetic Construction: Standing Wave* by Gabo, as well as the writings of movement and duration by Bergson. Like Bergson, Moholy-Nagy and Gabo were primarily interested in the subjective perception of time. While Bergson argued that mechanical representation distorts the experience and integrity of duration, I have argued that through their works Moholy-Nagy and Gabo orchestrate a multitude of movement rhythms from mechanical media to explore new perceptions and spectacles of duration. In this regard, there are intersecting approaches to the perception of movement and time between each artist and

Bergson's philosophies, and also key differences in expressing and conceptualising duration through visual media.

These two artists are important for understanding how contemporary art history frames kineticisim to contribute to a consciousness of time specific to contemporaneity today. Furthermore, the modes of peculiar time that build through my emphasis of Bergson's duration in relation to these artists is a conflicted sense of duration: one where new perceptions and sensations of time are orchestrated by repetition, representation, and movement of mechanical media. Bergson's concept of duration is useful for considering time as a constant transformation that preserves the past within it. For Moholy-Nagy and Gabo, movement in art is used to bring attention to temporality in a spectacle of modern mechanical sensations, and articulates folds of time within the present through Gabo's virtual volumes and Moholy-Nagy's modulation and refraction of light.

Finally, Eliasson's artwork *Your Negotiable Panorama* was addressed in Chap. 6 to highlight a disparity in Popper's interpretation of the term 'virtual'. Popper considers virtuality to describe an instantaneous, ephemeral, and immaterial exchange of simulation of reality. Like Moholy-Nagy and Gabo before him, Eliasson is concerned with the material and immaterial effects of movement, and is concerned with the virtual as something that is real, rather than a simulation of the real. Eliasson also orchestrates a consciousness of time that encourages the viewer to bring attention to their perception of movement in time. Eliasson draws from a Bergsonian understanding of the present as a site of differentiation from the virtual to the actual, while also preserving the past within it. Unlike Popper, the virtual for Eliasson becomes actualised and produces materiality. I also elaborated on the crystalline as a way to describe moments of perceiving time as a series of folds of present, past, and future temporalities when viewing the installation.

Throughout the book I approached the criticisms that posit kinetic art as an unrequited, obsolete, forgotten cabinet of curiosities as a wider reflection of the changing reactions to the relationship between time and art. Each chapter therefore focused on a polemic within the contemporary art history of kinetic sculpture: duration, systems aesthetics, demateriality, and the virtual. Each polemic was considered as a key contributing factor for considering avant-garde kinetic art as an antiquated, mechanical, material, formal, and a precursor to digital, conceptual, and participatory practices in art. That is, each of these polemics have contributed to Pierre's argument that there is a clear absence of critical historical discussion around the impact of kinetic art in contemporary society (p. 91). Identifying these polemics aided in identifying the roles and effects of kinetic art history that contribute to contemporary perspectives of contemporaneity.

Throughout this book the approach to contemporary art history was largely influenced by Smith's method for discussing contemporary art in *What is Contemporary Art?* and *Contemporary Art: World Currents*. In these texts Smith identifies three intersecting currents that have developed in art since the 1990s, one of which was constructive in this book for unpacking the reasons as to why there has been a focus on avant-garde kinetic artists in recent large-scale exhibitions, as well as a discussion around contemporaneity in contemporary art. The discernible and interweaving tendency of artists, institutions, and criticism in the present day is the inclination to resensationalise key tropes of modern art: 'reflexivity and avant-garde experimentality' (2009, p. 256). Smith's approach to defining and discussing contemporary art is useful for reconsidering the roles and functions of avant-garde and contemporary kinetic sculpture and installation. In taking this approach, the book expanded on a variety of ways in which artists, historians, and curators have orchestrated and drawn from kineticism to 'resensationalise' or 're-modernise' contemporary understandings of temporality and contemporaneity. Rather than considering kinetic sculpture as a modern antiquated mechanical medium due to the emerging post-modern aesthetics in the 1970s, I argue that kinetic sculpture is a contemporary and historical practice that can be used to reflect on the nature of contemporaneity in art and society today.

As part of that argument, I addressed a number of areas where artists and curators facilitated a tendency of what Smith describes as 'resensationalisaiton' of modern tropes in art, which build a form of remodernism in contemporary art. While Smith elaborates on the renovations of art institutions such as the Tate in London and MoMA in New York, as well as the exhibitions of large-scale modern artworks at the DIA foundation, I applied identified patterns of remodernism in relation to kinetic sculpture. Specifically, the artists and exhibitions that I focused on pay attention to the perception and orchestration of temporality through actual movement. This builds a tendency to engage with kinetic works of art from a temporal, rather than a spatial perspective. Furthermore, the experience of being in real-time movement with the kinetic object has often been interpreted through Bergson's modern conception of duration that is, the time we experience subjectively.

For instance, the renewed interest in avant-garde kinetic art creates a resensationalisation of the practice for contemporary audiences. While also, as I argued, contemporary artists such as Kapoor and Eliasson remodernise key driving tropes of avant-garde art such as duration, temporality, and orchestrating perceptions of time through actual motion. It is here that this book makes a contribution to the field of contemporary art history, and is valuable for discussing the effects of remodernism while focusing on practices that Smith has not yet analysed. In light of the recent surge of exhibitions that focus on the role of time in art as well as kineticisim, this analysis will become more important to the fields of contemporary art history.

Throughout this book I present movement in art as an expression of temporality through the orchestration of actual movement of mechanical media. The conceptualisation of temporality that is often presented through movement in real time with the viewer is one that is universal, of continual becoming and differentiation between the future and present temporalities, and one that regards space as a coordinate of time. As Jameson identified, the rise of postmodern theory in the middle of the twentieth century worked to render temporality 'as non-person' (p. 695), a perspective which, facilitated a spatialisation of the hierarchical power structures of modernity. This outlook also aided an association of kinetic sculpture with modern antiquity precisely because of the strong link between kineticism as a means to express and explore the changing conceptions of temporality. The appropriation of general systems theory, as well as the rising popularity of digital media in art, and an approach to art that dematerialises the object from the work of art, reaffirmed a consideration for kinetic sculpture as a formal, discrete, and crafted spectacle of the machine age that in many ways excluded the term 'kinetic' from postmodern interpretation.

Additionally, Popper's consideration of digital art as a virtual, immaterial, conceptual art or the 'more refined' heir to early mechanical and technological art such as kinetic sculpture, has received limited critical engagement in the field of media art history. My focus on Popper's use of concepts such as demateriality and virtuality to perpetuate and justify a post-kinetic framework for media art considers new pathways for engaging with and reflecting on the role of actual movement in contemporary art experiences. While Popper has argued that a kineticism is an aspect of media art history that generally moves from the materiality of kinetic movement to immaterial, virtual simulations of movement, I have argued that there are alternative avenues for discussing the roles and effects of kineticism in contemporary art history. Considering the strong influence Popper has within the fields of media art history, technological art, and participatory art, the research in this book focuses on only a number of concepts addressed by Popper. Future reflections on Popper's influence in contemporary media art history and producing new art historical lineages that intersect, contribute to, or run parallel to Popper's framework, would therefore benefit from the method and findings within this book.

My approach to kinetic art as a mechanism for understanding the facets of contemporaneity in art and society today is therefore also useful for the scholarship that discusses real-time art, time-based art, and movement studies within the wider scope of contemporary art history. An endemic problem with studying movement in art is that the field of scholarship and artistic practices are incredibly open, particularly with the present day popularity of installation among contemporary artists. The orchestration and study of movement in art is a prominent feature in digital, installation, participatory, interactive, and conceptual art, as well as dance, architecture, film, photography, expanded cinema, and theatre.

While this open applicability of the term 'kinetic' has been, at times, a problem when discerning the breadth of relevant literature and the structure of the book, this also means that the approaches, arguments, and works addressed in this book are relevant to a broad variety of intersecting fields of contemporary art and analysis. Throughout this book I highlighted kinetic sculpture as a contemporary preoccupation that reflects on Popper and Burnham's influential approaches to kinetic art, and present further avenues for new discussing of movement in relation to contemporary art that intersect with, and at times, run parallel to, their criticisms. More specifically, this approach will be most useful for those interested in examining the role and function of time in art in contemporary society, as well as understanding modernity as, in the words of Habermas, a 'system that is endless in nature'.

For scholarship that focuses on contemporary art and a Bergson-Deleuzian philosophy, there is further research that could focus on aspects of the role of affect in art, the relationship between the virtual and the actual in time, perception, duration, and Deleuze's movement-image, time-image, and the crystalline. In this book I applied and discussed these philosophical terms through an interpretation of the use of actual movement in art and, because of this, my argument cleaves a passage for Bergson and Deleuze to be used to interpret kinetic art in new ways. Such an approach points to further aspects that could be explored through the practical implications of the philosophical concepts of duration, the virtual, and the crystalline regime of images as a means for drawing out the affects of artistic practices such as kinetic art.

As Smith has articulated in *What is Contemporary Art?* that '[c]ontemporaneity itself has many histories, and histories within the histories of art' (p. 6). While I have argued throughout this book that the effects of kineticism in art and contemporary art history contributes to a consciousness of contemporaneity today, this approach to contemporaneity is one of many ways to understand the *peculiar time* that is unique to society today. My approach to avant-garde kinetic art as a resource for understanding the expressions and conceptualisations of time today therefore has the potential to intersect with new understandings of contemporaneity in the future.

As demonstrated by the artists, works, and exhibitions that focus on actual movement in art, the role of kinetic art in contemporary art history does not solely function as a mechanical precursor to contemporary art and electronic media. Instead, throughout this book I have encouraged a reconsideration of how contemporary and historical uses of movement and time in art can be considered to intersect with, and run parallel to, the genealogy of digital media in art. Therefore, this book also invites approaches to art that seek to destabilise binary distinctions between digital and analogue media, as well as examine the material and immaterial effects of movement in art as part of a consideration of the nature of contemporaneity in contemporary art and society.

References

Althusser L, Balibar É (1970) Reading capital, (trans. Brewster B.). NLB, New York

Bertalanffy L (1968) General systems theory; foundations, developments applications. G. Braziller, New York

Burnham J (1968) Systems esthetics. Artforum 7(1):30–35

Christov-Bakargiev C (2008) 'Revolutions forms that Turn: the Impulse to Revolt', 2008 Sydney Biennale: Revolutions—Forms That Turn. Thames and Hudson, Sydney

Danchev A (ed) (2011) 100 Artists' Manifestos; from the futurists to the stuckists. Penguin, London

Habermas J (1990) The philosophical discourse of modernity, 4th edn. (trans. Lawrence F.). MIT Press, Cambridge

Jameson F (2003) The end of temporality. Crit Inq 29(4):695–718

Lee P (2004) Chronophobia: on time in the art of the 1960s. MIT Press, Cambridge

Pierre A (2010) Instability: the visual/bodily perception of space in kinetic environments'. The 'do-it-yourself artwork: Participation From Fluxus to New Media. Manchester University Press, Manchester
Popper F (1968) Origins and developments of kinetic art. New York Graphic Society, New York
Popper F (1975) Art, action and participation. Studio Vista, London
Popper F (2007) From technological to virtual art. MIT Press, Cambridge.
Smith T (2009) What is contemporary art? University of Chicago Press, Chicago
Smith T (2011) Contemporary art: world currents. Prentice Hall, Upper Saddle River

Index

© Springer Nature Singapore Pte Ltd. 2017

C. Chau, *Movement, Time, Technology, and Art*, Springer Series on Cultural Computing, DOI 10.1007/978-981-10-4705-3

FSC
www.fsc.org
MIX
Papier | Fördert gute Waldnutzung
FSC® C083411

Kolibri 360 GmbH
Ferdinand-Jühlke-Straße 7
99095 Erfurt, Deutschland
produktsicherheit@kolibri360.de